Introduction to HILBERT SPACE

AND THE THEORY OF SPECTRAL MULTIPLICITY

PAUL R. HALMOS

SECOND EDITION

AMS CHELSEA PUBLISHING
American Mathematical Society • Providence, Rhode Island

Library of Congress Catalog Card Number 57-12834
ISBN 0-8218-1378-1

Printed in the United States of America.

∞ The paper used in this book is acid-free and falls within the guidelines
established to ensure permanence and durability.

10 9 8 7 6 5 4 3 2 1   02 01 00 99 98

PREFACE

My main reason for writing this book was to make available to English-speaking students the results of Chapter III, the so-called multiplicity theory. The only exposition of that theory that has been easily available in America is the one given by Stone, who discussed self-adjoint operators on a separable Hilbert space. The theory as I present it deals with arbitrary spectral measures and includes, consequently, the multiplicity theory of (bounded or unbounded) normal operators on a not necessarily separable Hilbert space, and includes, as another useful special case, the multiplicity theory of unitary representations of locally compact abelian groups. In view of the fact that a weakly closed, self-adjoint, commutative operator algebra has a lot of projections in it, the structure theory for Boolean algebras of projections, as developed in Chapter III, applies to such operator algebras also.

I have been fortunate in being able to make use of several simplifications of Hilbert space theory, some of which were published only in the last five years. As examples of such recent contributions I mention Eberlein's proof of the spectral theorem and the detailed treatment of the multiplicity theory by Plessner and Rohlin. The work of the latter authors, in turn, is obviously very strongly influenced by the pioneering research of Wecken. The approach to multiplicity theory which I present has some claim to novelty, but in its fundamental ideas it is essentially a permutation of what I learned from Wecken and from Nakano.

The first two chapters of the book are not new at all and they are there only to prepare the way for Chapter III. The last clause is not, however, to be taken literally—one can draw a shorter and straighter line between the axioms of Hilbert space and the theory of multiplicity than the one I have drawn. Such material as does not directly contribute to Chapter III has the purpose of nailing down the edges, so to speak—of supporting the strictly necessary material by illuminating and illustrating it. Despite the presence of "irrelevant" theorems, large parts of the theory of Hilbert space are still conspicuous by their absence: I do not define unbounded operators, for instance, and I do not even mention any of the several valuable applications of the theory to interesting special cases.

There are three technical details that the reader should know. (1) Since some of the notation which is used throughout the book is established in §0, both the expert and the beginner are advised to glance at that initial section. (2) There are a few statements, printed formally as theorems, which are not supported by even one word of proof. They exist for purposes of reference and they are not proved, because I considered them trivial. (3) The reference system is simple and standard. An expression such as $u.v$, where u and v are ordinal numbers, refers to Theorem v in §u.

In conclusion I want to express my warmest thanks to Arlen Brown, M. Gerstenhaber, M. M. Gutterman, and E. A. Michael for their aid in preparing this book. They read the manuscript, made many valuable suggestions, and would not back down when I objected to their criticism. I am also grateful to my colleagues Irving Kaplansky and I. E. Segal for several stimulating conversations about multiplicity theory.

P. R. H.

CONTENTS

CHAPTER III

THE ANALYSIS OF SPECTRAL MEASURES

§0. Prerequisites and Notation

The principal prerequisite for an intelligent reading of this little book is a thorough knowledge of what is usually called the theory of functions of a real variable. We use that phrase, as it is always used, to denote a hodge-podge of the theories of sets, cardinal numbers, topological (and particularly metric) spaces, measure, and integration. References for results used but not developed in the text (as well as occasional references to the sources of our material and to detailed presentations of subjects we shall barely have time to mention) are to be found at the end of the book.

We devote the remainder of this section to a detailed description of our terminological and notational conventions and to the statement of a representation theorem for linear functionals which we need in a form slightly different from the one in which it is usually given.

The word *family* is used throughout (as a generalization of *sequence*) to denote an indexed set, so that, for instance, a family $\{\alpha_j\}$ of real numbers is a real-valued function on a certain index set $\{j\}$. Any adjective (such as finite or countable), when applied to a family, is to be interpreted so as to modify the index set which serves as the domain of that family. If $\{\alpha_j\}$ is a family of objects, each object α_j is called a *term* of the family.

The symbol δ_{jk} is the Kronecker delta: its value is 1 or 0 according as $j = k$ or $j \neq k$. The symbol $\aleph_0$ denotes the cardinal number of the set of all integers. The letter χ (almost always used with a subscript) is reserved for characteristic functions, so that, for instance, if M is a subset of a space X and if t is a point of X, then $\chi_M(t) = 1$ or 0 according as t does or does not belong to M.

The word *polynomial* without an adjective means a polynomial with complex coefficients; the modification in the phrase *real polynomial* indicates a polynomial with real coefficients. The complex conjugate of a complex number α is denoted by α^*. The least upper bound and the greatest lower bound of a set M of real numbers are denoted by symbols such as $\sup\{\alpha:\alpha \in M\}$ and $\inf\{\alpha:\alpha \in M\}$ respectively.

The empty set is denoted by 0. The symbol $\{\cdots:\cdots\}$ is used for "the set where...", so that, for instance, $\{\alpha:\alpha > 0\}$ is the set of all positive real numbers. The symbols $\cup$, $\cap$, $-$, and $\subset$ are used for union, intersection, relative complement, and not necessarily proper set inclusion respectively. The symbol ϵ is used to indicate the belonging of an element to a set; the negation of a belonging assertion is indicated by a similar use of ϵ'. The Cartesian product of two sets M and N is denoted by $M \times N$.

The convergence of a sequence $\{x_n\}$ of points in a metric space to a point x is denoted by $x_n \to x$. The closure of a subset M of a metric (or, more generally, of a topological) space is denoted by $\bar{M}$.

By a *measure* (without adjectives) we shall always mean a non-negative and countably additive set function μ defined on a Boolean σ-algebra $\mathbf{S}$ of subsets of a set X. Almost all the measures we shall encounter will be *finite measures*, i.e. such that $\mu(X) < \infty$. A *complex measure* is a complex-valued, countably additive set function. Since the real and imaginary parts of a complex measure are countably additive, and since, therefore, each of these parts is the difference of two measures, it makes sense to integrate with respect to a complex measure; the process is to be carried out by expressing the given complex measure as a linear combination of measures, as just indicated, and then forming the corresponding linear combination of ordinary integrals.

If $(X, \mathbf{S}, \mu)$ is a measure space, if M is a measurable subset of X (i.e. if $M \subset X$ and $M \epsilon \mathbf{S}$), and if a (complex-valued) function f is integrable with respect to μ on M, then the value of the integral is denoted by $\int_M f(t)\, d\mu(t)$ or $\int_M f\, d\mu$; if $M = X$, the subscript is omitted.

If μ is a measure and if α is a positive number, the set of all complex-valued measurable functions f such that $|f|^\alpha$ is integrable with respect to μ is denoted by $\mathfrak{L}_\alpha(\mu)$. (The only values of α which will interest us are 1 and 2.) If two functions in $\mathfrak{L}_\alpha(\mu)$ differ only on a set of measure zero, they are regarded as identical.

A useful prepositional distinction is made by saying that a measure μ is defined *on* $\mathbf{S}$ and *in* X. This usage may be extended slightly. If X is a set, if $\mathbf{S}$ is a Boolean σ-algebra of subsets of X, and if M is a set in $\mathbf{S}$, we shall speak of a measure μ defined *in* M, meaning that μ is defined *on* $\mathbf{S}$ and $\mu(X - M) = 0$.

The representation theorem that we mentioned earlier may be stated as follows. Suppose that L is a complex-valued function whose domain is the set of all real polynomials (in one variable) and which is such that $L(\alpha p + \beta q) = \alpha L(p) + \beta L(q)$ whenever α and β are real numbers and p and q are real polynomials—suppose, in other words, that L is a linear functional of polynomials. Let X be the real line and let Λ be a compact subset of X; for any complex-valued, bounded function f on Λ write $\mathbf{N}_\Lambda(f) = \sup \{|f(\lambda)|:\lambda \in \Lambda\}$. If the linear functional L is *bounded* in the sense that there exists a positive real number α such that $|L(p)| \leqq \alpha\mathbf{N}_\Lambda(p)$ for all real polynomials p, then there exists a unique complex measure μ defined *on* the class of all Borel subsets of X and *in* Λ and such that $L(p) = \int_\Lambda p\, d\mu$ for all real polynomials p. The complex measure μ has, moreover, the property that $|\mu(M)| \leqq \alpha$ for every Borel subset M of X.

If that were all, it would be bad enough—but we need even more. The more that we need is the extension of the theorem to two dimensions. The statement of the more general result is very easy to describe: it is obtained from the statement above by changing the parenthetical phrase "in one variable" to "in two variables," and interpreting the symbol X as the Cartesian product of two real lines (or, equivalently, as the complex plane).

CHAPTER I

THE GEOMETRY OF HILBERT SPACE

§1. Linear Functionals

Throughout this book we shall work with vector spaces over the field of complex numbers, or, as they may be more briefly described, complex vector spaces. The simplest and yet by far the most important example of a complex vector space is the set $\mathfrak{C}$ of all complex numbers, with the vector operations of addition and scalar multiplication interpreted as the ordinary arithmetic operations of addition and multiplication of complex numbers.

We recall an elementary definition. A *linear transformation* from a complex vector space $\mathfrak{H}$ to a complex vector space $\mathfrak{K}$ is a mapping A from $\mathfrak{H}$ into $\mathfrak{K}$ such that $A(\alpha x + \beta y) = \alpha A x + \beta A y$ identically for all complex numbers α and β and all vectors x and y in $\mathfrak{H}$. Just as the special vector space $\mathfrak{C}$ plays a distinguished role among all complex vector spaces, similarly linear transformations whose range space $\mathfrak{K}$ coincides with $\mathfrak{C}$ (such linear transformations are called *linear functionals*) play a distinguished role among all linear transformations. Explicitly: a linear functional on a complex vector space $\mathfrak{H}$ is a complex-valued function ξ on $\mathfrak{H}$ such that (and now we proceed, for the sake of variety, to state the definition of linearity in terms slightly different from the ones used above)

(i) ξ is *additive* (i.e. $\xi(x + y) = \xi(x) + \xi(y)$ for every pair of vectors x and y in $\mathfrak{H}$), and

(ii) ξ is *homogeneous* (i.e. $\xi(\alpha x) = \alpha\xi(x)$ for every complex number α and for every vector x in $\mathfrak{H}$).

It is sometimes convenient to consider, along with linear functionals, the closely related *conjugate linear functionals* whose definition differs from the one just given in that the equation $\xi(\alpha x) = \alpha\xi(x)$ is replaced by $\xi(\alpha x) = \alpha^*\xi(x)$. There is a simple and obvious relation between the two concepts: a necessary and sufficient condition that a complex-valued function ξ on a complex vector space be a linear functional is that ξ^* be a conjugate linear functional.

§2. Bilinear Functionals

For the theory that we shall develop, the concept of a bilinear functional is even more important than that of a linear functional. A *bilinear functional* on a complex vector space $\mathfrak{H}$ is a complex-valued function φ on the Cartesian product of $\mathfrak{H}$ with itself such that if $\xi_y(x) = \eta_x(y) = \varphi(x, y)$, then, for every x and y in $\mathfrak{H}$, ξ_y is a linear functional and η_x is a conjugate linear functional.

This definition of a bilinear functional is different from the one commonly used in the theory of vector spaces over an arbitrary field; the usual definition requires that, for every x and y in $\mathfrak{H}$, both η_x and ξ_y shall be linear functionals. An example of a bilinear functional in this "usual" sense may be manufactured by starting with two arbitrary linear functionals ξ and η and writing $\varphi(x, y) = \xi(x)\eta(y)$; an obviously related example of a bilinear functional in the sense in which we defined that concept is obtained by writing $\varphi(x, y) = \xi(x)\eta^*(y)$. The objects that we defined are sometimes called *Hermitian* bilinear functionals. Further examples of either usual or Hermitian bilinear functionals may be constructed by forming finite linear combinations of examples of the product type described above. After this brief comment on the peculiarity of our terminology (adopted for reasons of simplicity), we shall consistently stick to the definition that was formally given in the preceding paragraph.

It is easy to verify that if φ is a bilinear functional and if the function ψ is defined by $\psi(x, y) = \varphi^*(y, x)$, then ψ is a bilinear functional. A bilinear functional φ is *symmetric* if $\varphi = \psi$, or, explicitly, if $\varphi(x, y) = \varphi^*(y, x)$ for every pair of vectors x and y. A bilinear functional φ is *positive* if $\varphi(x, x) \geqq 0$ for every vector x; we shall say that φ is *strictly positive* if $\varphi(x, x) > 0$ whenever $x \neq 0$.

§3. Quadratic Forms

The *quadratic form* $\hat{\varphi}$ induced by a bilinear functional φ on a complex vector space is the function defined for each vector x by $\hat{\varphi}(x) = \varphi(x, x)$. Using this language and notation, we may paraphrase one of the definitions in the last paragraph of the preceding section as follows: φ is positive if and only if $\hat{\varphi}$ is positive in the ordinary sense of taking only positive values.

A routine computation yields the following useful result.

THEOREM 1. *If $\hat{\varphi}$ is the quadratic form induced by a bilinear functional φ on a complex vector space $\mathfrak{H}$, then*

$$\varphi(x, y) = \hat{\varphi}(\tfrac{1}{2}(x + y)) - \hat{\varphi}(\tfrac{1}{2}(x - y)) + i\hat{\varphi}(\tfrac{1}{2}(x + iy)) - i\hat{\varphi}(\tfrac{1}{2}(x - iy))$$

for every pair of vectors x and y in $\mathfrak{H}$.

The process of calculating the values of the bilinear functional φ from the values of the quadratic form $\hat{\varphi}$, in accordance with the identity in Theorem 1, is known as *polarization*. As an immediate corollary of this process we obtain (and we state in Theorem 2) the fact that a bilinear functional is uniquely determined by its quadratic form.

THEOREM 2. *If two bilinear functionals φ and ψ are such that $\hat{\varphi} = \hat{\psi}$, then $\varphi = \psi$.*

Theorem 2 in turn may be applied to yield a simple characterization of symmetric bilinear functionals.

THEOREM 3. *A bilinear functional φ is symmetric if and only if $\hat{\varphi}$ is real.*

Proof. If φ is symmetric, then $\hat{\varphi}(x) = \varphi(x, x) = \varphi^*(x, x) = \hat{\varphi}^*(x)$ for all x. If, conversely, $\hat{\varphi}$ is real, then the bilinear functional ψ, defined by $\psi(x, y) = \varphi^*(y, x)$, and the bilinear functional φ are such that $\hat{\varphi} = \hat{\psi}$; it follows from Theorem 2 that $\varphi = \psi$.

§4. Inner Product and Norm

An *inner product* in a complex vector space $\mathfrak{H}$ is a strictly positive, symmetric, bilinear functional on $\mathfrak{H}$. An *inner product space* is a complex vector space $\mathfrak{H}$ and an inner product in $\mathfrak{H}$. The vector space $\mathfrak{C}$ of all complex numbers becomes an inner product space if the inner product of α and β is defined to be $\alpha\beta^*$; in what follows we shall always interpret the symbol $\mathfrak{C}$, not merely as a vector space, but as an inner product space with this particular inner product.

It is convenient and, as it turns out, not confusing to use the same notation for inner product in all inner product spaces; the value of the inner product at an ordered pair of vectors x and y will be denoted by (x, y). The quadratic form induced by the inner product also has a universal symbol: its value at a vector x will be denoted by $\| x \|^2$. The positive square root $\| x \|$ of $\| x \|^2$ is called the *norm* of the vector x. Note that the norm of a vector α in the inner product space $\mathfrak{C}$ coincides with the absolute value of the complex number α.

Throughout this book, unless in some special context we explicitly indicate otherwise, the symbol $\mathfrak{H}$ will denote a fixed inner product space; all apparently homeless vectors will be presumed to belong to $\mathfrak{H}$ and the definitions of all concepts and the proofs of all theorems will pertain to $\mathfrak{H}$.

THEOREM 1. *A necessary and sufficient condition that $x = 0$ is that $(x, y) = 0$ for all y.*

Proof. If $(x, y) = 0$ for all y, then, in particular, $(x, x) = 0$ and consequently, since the inner product is strictly positive, $x = 0$. If, conversely, $x = 0$, then $(x, y) = (0x, y) = 0(x, y) = 0$. (Note that the proof of the converse is nothing more than the proof of the fact that if ξ is any linear functional, then $\xi(0) = 0$. It follows, of course, that if φ is any bilinear functional, then $\varphi(0, y) = \varphi(x, 0) = 0$ for all x and y.)

THEOREM 2. (*The parallelogram law.*) *For any vectors x and y,*

$$\| x + y \|^2 + \| x - y \|^2 = 2\| x \|^2 + 2\| y \|^2.$$

Proof. Compute.

The reader should realize the relation between Theorem 2 and the assertion that the sum of the squares of the two diagonals of a parallelogram is equal to the sum of the squares of its four sides.

The most important relation between vectors of an inner product space is orthogonality; we shall say that x is *orthogonal* to y, in symbols $x \perp y$, if $(x, y) = 0$. In terms of this concept Theorem 1 says that the only vector orthogonal to every vector is 0. For orthogonal vectors the statement of the parallelogram law may be considerably sharpened.

THEOREM 3. (*The Pythagorean theorem.*) *If $x \perp y$, then*

$$\| x + y \|^2 = \| x \|^2 + \| y \|^2.$$

The reader should realize the relation between Theorem 3 and the assertion that the square of the hypotenuse of a right triangle is the sum of the squares of its two perpendicular sides.

A family $\{x_j\}$ of vectors is an *orthogonal family* if $x_j \perp x_k$ whenever $j \neq k$. We shall have no qualms about using the obvious inductive generalization of the Pythagorean theorem, i.e. the assertion that if $\{x_j\}$ is a finite orthogonal family, then $\| \Sigma_j x_j \|^2 = \Sigma_j \| x_j \|^2$.

§5. The Inequalities of Bessel and Schwarz

A vector x is *normalized*, or is a *unit vector*, if $\| x \| = 1$; the process of replacing a non-zero vector x by the unit vector $x/\| x \|$ is called *normalization*. A family $\{x_j\}$ of vectors is an *orthonormal family* if it is an orthogonal family and each vector x_j is normalized, or, more explicitly, if $(x_j, x_k) = \delta_{jk}$ for all j and k.

THEOREM 1. *(Bessel's inequality.) If $\{x_j\}$ is a finite orthonormal family of vectors, then*

$$\Sigma_j |(x, x_j)|^2 \leqq \|x\|^2$$

for every vector x.

Proof.

$$0 \leqq \|x - \Sigma_j(x, x_j)x_j\|^2 = \|x\|^2 - \Sigma_j(x, x_j)(x_j, x) - \Sigma_j(x, x_j)^*(x, x_j)$$
$$+ \Sigma_j\Sigma_k(x, x_j)(x, x_k)^*(x_j, x_k) = \|x\|^2 - \Sigma_j |(x, x_j)|^2.$$

(The expressions (x, x_j) will occur frequently in our work; they are called the *Fourier coefficients* of the vector x with respect to the orthonormal family $\{x_j\}$.)

It is sometimes useful to realize that the strict positiveness of the inner product is not needed to prove the Bessel inequality. In the presence of strict positiveness, however, the statement of Bessel's inequality can be improved by adding to it the assertion that equality holds if and only if x is a linear combination of the x_j's. The proof of this addition is an almost immediate consequence of the observation that in the proof of Bessel's inequality there is only one place at which an inequality sign occurs.

THEOREM 2. *(Schwarz's inequality.)* $|(x, y)| \leqq \|x\| \cdot \|y\|$.

Proof. If $y = 0$, the result is obvious. If $y \neq 0$, write $y_0 = y/\|y\|$; since $\|y_0\| = 1$, i.e. since the family consisting of the one term y_0 is an orthononormal family, it follows from Bessel's inequality that $|(x, y_0)| \leqq \|x\|$.

Schwarz's inequality, just as Bessel's inequality, would be true even if the inner product were not strictly positive (but merely positive). Our proof of Schwarz's inequality is not delicate enough to yield this improvement: we made use of strict positiveness through the possibility of normalizing any non-zero vector. In the presence of strict positiveness, however, the statement of Schwarz's inequality can be improved by adding to it the assertion that equality holds if and only if x and y are linearly dependent; the proof of this addition is, in one direction, trivial and, in the other direction, a consequence of the corresponding facts about Bessel's inequality.

The Schwarz inequality has an interesting generalization. If $\{x_j\}$ is a non-empty, finite family of vectors, and if $\gamma_{jk} = (x_j, x_k)$, then the determinant of the matrix $[\gamma_{jk}]$ is non-negative; it vanishes if and only if the x_j's are linearly dependent.

§6. Hilbert Space

THEOREM 1. *The norm in an inner product space is*

strictly positive (*i.e.* $\| x \| > 0$ *whenever* $x \neq 0$),
positively homogeneous (*i.e.* $\| \alpha x \| = | \alpha | \cdot \| x \|$), *and*
subadditive (*i.e.* $\| x + y \| \leqq \| x \| + \| y \|$).

Proof. The strict positiveness of the norm is merely a restatement of the strict positiveness of the inner product. The positive homogeneity of the norm is a consequence of the identity

$$\| \alpha x \|^2 = (\alpha x, \alpha x) = \alpha\alpha^*(x, x) = | \alpha |^2 \cdot \| x \|^2.$$

The subadditivity of the norm follows, using Schwarz's inequality, from the relations

$$\| x + y \|^2 = (x + y, x + y) \leqq \| x \|^2 + |(x, y)| + |(y, x)| + \| y \|^2$$
$$\leqq \| x \|^2 + 2 \| x \| \cdot \| y \| + \| y \|^2 = (\| x \| + \| y \|)^2.$$

THEOREM 2. *If the distance from a vector x to a vector y is defined to be $\| x - y \|$, then, with respect to this distance function, $\mathfrak{H}$ is a metric space.*

Proof. The fact that the distance function is *strictly positive* (i.e. that $\| x - y \| \geqq 0$, with equality holding if and only if $x = y$) follows from the strict positiveness of the norm. The fact that the distance function is *symmetric* (i.e. that $\| x - y \| = \| y - x \|$ for every pair of vectors x and y) follows from the positive homogeneity of the norm and the identity $(x - y) = (-1)(y - x)$. The validity of the *triangle inequality* (i.e. the relation $\| x - y \| \leqq \| x - z \| + \| z - y \|$ for every triple of vectors x, y, and z) follows from the subadditivity of the norm and the identity $x - y = (x - z) + (z - y)$.

In view of Theorem 2 we shall feel free to use, for inner product spaces, all such topological concepts as convergence, continuity, separability, dense set, closed set, and the closure of a set, and all such metric concepts as uniform continuity, Cauchy sequences, and completeness. We shall, in particular, need to make use of the continuity of the four operations (scalar multiplication, addition, and the formation of inner products and norms) which are intrinsic to inner product spaces.

THEOREM 3. *If* $\Phi_\alpha(x) = \alpha x$, $\Phi^+(x, y) = x + y$, $\Phi_y(x) = (x, y)$, *and* $\Phi(x) = \| x \|$ *whenever* α *is a complex number and* x *and* y *are vectors, then all the functions* Φ_α, Φ^+, Φ_y, *and* Φ *are uniformly continuous functions of all their arguments.*

Proof. The four assertions are consequences, respectively, of the following four inequalities.

$$\| \alpha x_1 - \alpha x_2 \| \leqq | \alpha | \cdot \| x_1 - x_2 \| .$$

$$\|(x_1 + y_1) - (x_2 + y_2)\| \leqq \| x_1 - x_2 \| + \| y_1 - y_2 \| .$$

$$|(x_1 , y) - (x_2 , y)| \leqq \| x_1 - x_2 \| \cdot \| y \| .$$

$$| \, \| x_1 \| - \| x_2 \| \, | \leqq \| x_1 - x_2 \| .$$

A *Hilbert space* is an inner product space which, as a metric space, is complete. It is worth noting that the special inner product space $\mathfrak{C}$ (cf. §§1 and 4) is in fact a Hilbert space. We extend the convention established in §4 by requiring that, from now on, *the inner product space* $\mathfrak{H}$ *under consideration shall in fact be a Hilbert space.*

A *normed* vector space is a vector space with a strictly positive, positively homogeneous, and subadditive norm; a *Banach space* is a normed vector space which, as a metric space, is complete. A small fraction of our results will be valid for Banach spaces as well as for the special Banach spaces (i.e. Hilbert spaces) that we are studying; whenever it is possible and convenient to do so, we shall arrange our proofs so that they make sense in any Banach space. The precise extent to which Hilbert spaces differ from general Banach spaces has received quite a bit of attention; it may be expressed by saying that the norm in a Hilbert space is essentially quadratic in character, in the sense, for instance, that the parallelogram law is valid.

The completeness of Hilbert space is, to be sure, an essential part of its structure, but it is unessential in the sense that an inner product space can always be completed to be a Hilbert space. More precisely it is true that the linear operations and the inner product may be uniquely extended to the ordinary metric completion of an inner product space so that the completion becomes a Hilbert space.

§7. Infinite Sums

A family $\{x_j\}$ of vectors will be called *summable* with *sum* x, in symbols $\Sigma_j x_j = x$, if for every positive number ε there exists a finite set J_0 of indices such that $\| x - \Sigma_{j\epsilon J} x_j \| < \varepsilon$ whenever J is a finite set of indices containing J_0. It is clear from this definition that a finite family of vectors is always summable and that its sum, in the present sense, coincides with the elementary concept of vector sum. As another example we mention the fact (whose proof is a not particularly difficult exercise) that a sequence $\{\alpha_n\}$ of vectors in the Hilbert space $\mathfrak{C}$ is summable with sum α if and only if the ordinary numerical series $\Sigma_{n=1}^{\infty} \alpha_n$ is absolutely convergent to the value α. We emphasize the fact (and we shall make use of it below) that our definition makes sense, in particular, in the space $\mathfrak{C}$ and hence that such relations as $\Sigma_j \alpha_j = \alpha$ are meaningful (though not necessarily true) for not necessarily countable families $\{\alpha_j\}$ of complex numbers.

It follows from the last remark that the theorem which we have been calling Bessel's inequality makes sense for not necessarily finite (nor even necessarily countable) orthonormal families. It not only makes sense—it is true. The proof requires nothing more than the observation that, by the definition of sums, it is sufficient to consider finite families. We propose, accordingly, to change our custom and, in the sequel, when we refer to Bessel's inequality, to have in mind the generalization just now discussed. More formally: *Bessel's inequality* is to be interpreted as the theorem obtained from 5.1 by deleting the word "finite." One amusing consequence of the Bessel inequality in this form is the proposition that if $\{x_n\}$ is an orthonormal sequence, then $(x, x_n) \to 0$ for every vector x, i.e. that the Fourier coefficients of x tend to 0.

THEOREM 1. *If $\Sigma_j x_j = x$, then $\Sigma_j \alpha x_j = \alpha x$ for every complex number α.*

THEOREM 2. *If $\{x_j\}$ and $\{y_j\}$ are two families of vectors, indexed by the same set $\{j\}$, and if $\Sigma_j x_j = x$ and $\Sigma_j y_j = y$, then $\Sigma_j(x_j + y_j) = x + y$.*

THEOREM 3. *If $\Sigma_j x_j = x$, then $\Sigma_j(x_j, y) = (x, y)$ and $\Sigma_j(y, x_j) = (y, x)$ for every vector y.*

The proofs of all three theorems are quite elementary; they are, in fact, consequences of the following three relations (valid for any finite set J of indices) respectively:

$$\| \alpha x - \Sigma_{j\epsilon J} \alpha x_j \| = | \alpha | \cdot \| x - \Sigma_{j\epsilon J} x_j \| ,$$

$$\|(x + y) - \Sigma_{j\epsilon J} (x_j + y_j)\| \leqq \| x - \Sigma_{j\epsilon J} x_j \| + \| y - \Sigma_{j\epsilon J} y_j \| ,$$

$$|(x, y) - \Sigma_{j\epsilon J} (x_j , y)| = |(x - \Sigma_{j\epsilon J} x_j , y)| \leqq \| x - \Sigma_{j\epsilon J} x_j \| \cdot \| y \| .$$

§8. Conditions for Summability

THEOREM 1. *A family $\{x_j\}$ of vectors is summable if and only if for every positive number ε there exists a finite set J_0 of indices such that $\| \Sigma_{j\epsilon J} x_j \| < \varepsilon$ whenever J is a finite set of indices disjoint from J_0. If $\{x_j\}$ is summable, then the set of those indices j for which $x_j \neq 0$ is countable.*

Proof. If $\{x_j\}$ is summable with sum x, then for every positive number ε there exists a finite set J_0 such that $\| x - \Sigma_{j\epsilon J} x_j \| < \frac{\varepsilon}{2}$ whenever $J \supset J_0$. It follows that if $J \cap J_0 = 0$, then

$$\| \Sigma_{j\epsilon J} x_j \| = \| \Sigma_{j\epsilon J \cup J_0} x_j - \Sigma_{j\epsilon J_0} x_j \| \leqq \| x - \Sigma_{j\epsilon J \cup J_0} x_j \| + \| x - \Sigma_{j\epsilon J_0} x_j \| < \varepsilon.$$

If, conversely, the condition is satisfied, then for every positive integer n there exists a finite set J_n such that $\| \Sigma_{j\epsilon J} x_j \| < \frac{1}{n}$ whenever $J \cap J_n = 0$. By replacing J_n by $J_1 \cup \cdots \cup J_n$, $n = 1, 2, \cdots$, we see that there is no loss of generality in assuming (and we do therefore assume) that the sequence $\{J_n\}$ of finite sets is increasing. (From these considerations we can already deduce the second assertion of the theorem. If, indeed, an index j does not belong to the countable set $J_1 \cup J_2 \cup \cdots$, then $\| x_j \| < \frac{1}{n}$ for every positive integer n and consequently $x_j = 0$.) To complete the proof of summability, note that if $n < m$, then

$$\| \Sigma_{j\epsilon J_m} x_j - \Sigma_{j\epsilon J_n} x_j \| = \| \Sigma_{j\epsilon J_m - J_n} x_j \| < \frac{1}{n},$$

since $(J_m - J_n) \cap J_n = 0$. It follows from the completeness of Hilbert space that there exists a vector x such that $\| \Sigma_{j\epsilon J_n} x_j - x \| \rightarrow 0$. If J is any finite set of indices containing J_n, then

$$\| x - \Sigma_{j\epsilon J} x_j \| \leqq \| x - \Sigma_{j\epsilon J_n} x_j \| + \| \Sigma_{j\epsilon J - J_n} x_j \|$$

and therefore $\{x_j\}$ is summable with sum x.

The second part of Theorem 1 asserts that our concept of summation is more of a notational convenience than a great generalization of the more elementary concept of infinite series.

THEOREM 2. *An orthogonal family $\{x_j\}$ of vectors is summable if and* *only if the family* $\{\|x_j\|^2\}$ *of positive numbers is summable; this condition may also be expressed by writing* $\Sigma_j \|x_j\|^2 < \infty$. *If* $x = \Sigma_j x_j$, *then* $\|x\|^2 = \Sigma_j \|x_j\|^2$.

Proof. If $\{x_j\}$ is summable, then for every positive number ε there exists a finite set J_0 such that $\|\Sigma_{j\epsilon J} x_j\| < \varepsilon$ whenever $J \cap J_0 = 0$, and consequently

$$\Sigma_{j\epsilon J} \|x_j\|^2 = \|\Sigma_{j\epsilon J} x_j\|^2 < \varepsilon^2$$

whenever $J \cap J_0 = 0$. If, conversely, $\Sigma_j \|x_j\|^2 < \infty$, then for every positive number ε there exists a finite set J_0 such that $\Sigma_{j\epsilon J} \|x_j\|^2 < \varepsilon^2$ (and consequently $\|\Sigma_{j\epsilon J} x_j\| < \varepsilon$) whenever $J \cap J_0 = 0$; summability follows from Theorem 1. The second assertion of the theorem is a consequence of the relations

$$(x, x) = (\Sigma_j x_j, x) = \Sigma_j (x_j, x) = \Sigma_j (x_j, \Sigma_k x_k)$$
$$= \Sigma_j \Sigma_k (x_j, x_k) = \Sigma_j (x_j, x_j).$$

(The last step in this chain of equations depends on the obvious fact that if all but one of the terms of a family of vectors, or, in particular, of complex numbers, are zero, then that family is summable and its sum is the exceptional term.)

Note that the second part of Theorem 2 is the obvious generalization of the Pythagorean theorem to not necessarily finite sums; just as in the case of Bessel's inequality we shall in the sequel use the phrase *Pythagorean theorem* to refer to the generalized version.

§9. Examples of Hilbert Spaces

A typical and general example of a Hilbert space is the space $\mathfrak{L}_2(\mu)$ of all complex-valued, measurable, and square-integrable functions on a measure space X with measure μ (with the usual understanding that two functions which differ on a set of measure zero only are to be identified). The linear operations in this space (as in every function space) are the usual pointwise operations and the inner product is defined by $(f, g) = \int f(t) g^*(t)\, d\mu(t)$.

An important special case of the example in the preceding paragraph is the one in which every subset of X is measurable and has as measure the number of its points. By an obvious change of notation (from $f(t)$ to ξ_j) the typical element of this Hilbert space becomes a family $\{\xi_j\}$ of complex numbers with the property that $\Sigma_j |\xi_j|^2 < \infty$; scalar multiplication, addition, and inner product are defined by

$$\alpha\{\xi_j\} = \{\alpha\xi_j\}, \qquad \{\xi_j\} + \{\eta_j\} = \{\xi_j + \eta_j\},$$

and

$$(\{\xi_j\}, \{\eta_j\}) = \Sigma_j \xi_j \eta_j^*,$$

respectively. (It is understood, of course, that the index set $\{j\}$ is the same for all vectors.)

An important generalization of the example in the preceding paragraph is obtained as follows. Let $\{\mathfrak{H}_j\}$ be a family of Hilbert spaces and denote by $\Sigma_j \mathfrak{H}_j$ the set of all families $\{x_j\}$ of vectors such that $x_j \in \mathfrak{H}_j$ for all j and such that $\Sigma_j \|x_j\|^2 < \infty$. If scalar multiplication, addition, and inner product are defined in $\Sigma_j \mathfrak{H}_j$ by

$$\alpha\{x_j\} = \{\alpha x_j\}, \qquad \{x_j\} + \{y_j\} = \{x_j + y_j\},$$

and

$$(\{x_j\}, \{y_j\}) = \Sigma_j (x_j, y_j),$$

respectively, then $\Sigma_j \mathfrak{H}_j$ becomes a Hilbert space. (The proof of this fact is a straightforward imitation of the proof that applies to the case in which $\mathfrak{H}_j = \mathfrak{C}$ for all j.) The space $\Sigma_j \mathfrak{H}_j$ is called the *external direct sum* or simply the *direct sum* of the family $\{\mathfrak{H}_j\}$ of Hilbert spaces.

Further examples of Hilbert spaces are: (i) the set of all those functions, defined and analytic in the interior of the unit circle in the complex plane, the square of whose absolute value is integrable with respect to planar Lebesgue measure, and (ii) the set of all functions almost periodic with exponent 2 in the sense of Besicovitch.

§10. Subspaces

A *linear manifold* is a non-empty subset $\mathfrak{M}$ of $\mathfrak{H}$ such that if x and y are in $\mathfrak{M}$, then $\alpha x + \beta y \in \mathfrak{M}$ for every pair of complex numbers α and β. A *subspace* is a closed linear manifold. The easiest examples of subspaces are the set $\mathfrak{O}$ containing 0 only and the entire space $\mathfrak{H}$. Note that a subspace of a Hilbert space is a Hilbert space and that therefore we may (and frequently shall) apply to subspaces any proposition we please, as long as it is true of all Hilbert spaces.

If μ is Lebesgue measure in the real line, then the following subsets of the Hilbert space $\mathfrak{L}_2(\mu)$ are all linear manifolds:

(i) the set of all those functions f in $\mathfrak{L}_2(\mu)$ for which $f(t) = f(-t)$ for (almost) every t;

(ii) the set of all those functions in $\mathfrak{L}_2(\mu)$ which vanish at (almost) every point of a certain measurable set;

(iii) the set of all (essentially) bounded functions in $\mathfrak{L}_2(\mu)$.

The first two of these sets are subspaces; the last one demonstrates the fact that there exist linear manifolds which are not closed. For another example of a linear manifold which is not a subspace consider the Hilbert space of all families $\{\xi_j\}$ of complex numbers such that $\Sigma_j | \xi_j |^2 < \infty$ (cf. §9) and the subset of all those families which have only a finite number of non-zero terms.

It is easy to see that the intersection of any family of subspaces is a subspace. It follows that it makes sense to define the subspace *spanned* by an arbitrary subset $\mathfrak{M}$ of $\mathfrak{H}$ (the *span* of $\mathfrak{M}$, in symbols $\vee \mathfrak{M}$) as the intersection of all subspaces containing $\mathfrak{M}$, or, equivalently, as the least subspace containing $\mathfrak{M}$.

THEOREM 1. *If $\mathfrak{M}$ is a non-empty subset of $\mathfrak{H}$ and if $\mathfrak{N}$ is the set of all finite linear combinations of elements of $\mathfrak{M}$, then $\mathfrak{N}$ is a linear manifold and $\vee \mathfrak{M} = \bar{\mathfrak{N}}$ (=the closure of $\mathfrak{N}$).*

Proof. It is clear that $\mathfrak{N}$ is a linear manifold and hence that $\bar{\mathfrak{N}}$ is a subspace; since $\mathfrak{M} \subset \bar{\mathfrak{N}}$, the minimal property of $\vee \mathfrak{M}$ implies that $\vee \mathfrak{M} \subset \bar{\mathfrak{N}}$. On the other hand the fact that $\vee \mathfrak{M}$ is a linear manifold implies that $\mathfrak{N} \subset \vee \mathfrak{M}$. Since $\vee \mathfrak{M}$ is closed, it follows that $\bar{\mathfrak{N}} \subset \vee \mathfrak{M}$.

If $\mathfrak{M}$ and $\mathfrak{N}$ are subspaces, we shall use the symbol $\mathfrak{M} \vee \mathfrak{N}$ for the subspace $\vee (\mathfrak{M} \cup \mathfrak{N})$; more generally, if $\{\mathfrak{M}_j\}$ is any family of subspaces then $\vee_j \mathfrak{M}_j$ will denote the subspace $\vee (\bigcup_j \mathfrak{M}_j)$. It follows from these definitions that $\mathfrak{M} \vee \mathfrak{N}$ is the least subspace containing both $\mathfrak{M}$ and $\mathfrak{N}$, and, more generally, that $\vee_j \mathfrak{M}_j$ is the least subspace containing every term of the family $\{\mathfrak{M}_j\}$.

The essential results of this section can be described rather simply in the language of lattice theory. The possibility of a lattice-theoretic formulation is based on the trite observation that the set of all subspaces of $\mathfrak{H}$ is a partially ordered set with respect to inclusion. The fact that for any family $\{\mathfrak{M}_j\}$ of subspaces there exists a greatest subspace $(\bigcap_j \mathfrak{M}_j)$ contained in them all and there exists a least subspace $(\vee_j \mathfrak{M}_j)$ containing them all may be expressed by saying that this partially ordered set is a complete lattice. While this lattice has many interesting properties, it is not in general so accomodating as to be distributive, nor even modular. It turns out, in fact, that the lattice of all subspaces of a Hilbert space is modular only in the familiar finite-dimensional cases, and that it is distributive only for the extremely trivial spaces whose dimension is 0 or 1.

§11. Vectors in and out of Subspaces

A vector x is *orthogonal* to a subset $\mathfrak{M}$ of $\mathfrak{H}$, in symbols $x \perp \mathfrak{M}$, if $x \perp y$ for all y in $\mathfrak{M}$. The purpose of this section is to obtain two results which our geometric intuition makes obvious and desirable. The first result is that the minimum of the distances from any fixed vector to the vectors of any fixed subspace is always attained; the second result is, essentially, that if a subspace is a proper subset of $\mathfrak{H}$, then there exists a non-zero vector orthogonal to the subspace.

THEOREM 1. *If $\mathfrak{M}$ is a subspace, if x is a vector, and if $\delta = \inf\{\|y - x\|: y \in \mathfrak{M}\}$, then there exists a vector y_0 in $\mathfrak{M}$ such that $\|y_0 - x\| = \delta$.*

Proof. Let $\{y_n\}$ be a sequence of vectors in $\mathfrak{M}$ such that $\|y_n - x\| \to \delta$. It follows from the parallelogram law that

$$\|y_n - y_m\|^2 = 2\|y_n - x\|^2 + 2\|y_m - x\|^2 - 4\|\tfrac{1}{2}(y_n + y_m) - x\|^2$$

for every n and m. Since $\frac{1}{2}(y_n + y_m) \in \mathfrak{M}$, it follows that

$$\|\tfrac{1}{2}(y_n + y_m) - x\|^2 \geqq \delta^2$$

and hence that

$$\|y_n - y_m\|^2 \leqq 2\|y_n - x\|^2 + 2\|y_m - x\|^2 - 4\delta^2.$$

As $n \to \infty$ and $m \to \infty$, the right side of the last written relation tends to $2\delta^2 + 2\delta^2 - 4\delta^2 = 0$, so that $\{y_n\}$ is a Cauchy sequence. If $y_n \to y_0$, then $y_0 \in \mathfrak{M}$ and, by the continuity of the norm,

$$\|y_0 - x\| = \lim_n \|y_n - x\| = \delta.$$

THEOREM 2. *If $\mathfrak{M}$ and $\mathfrak{N}$ are subspaces such that $\mathfrak{M} \subset \mathfrak{N}$ and $\mathfrak{M} \neq \mathfrak{N}$, then there exists a non-zero vector z in $\mathfrak{N}$ such that $z \perp \mathfrak{M}$.*

Proof. Let x be any vector in $\mathfrak{N}$ which is not in $\mathfrak{M}$ and write $\delta = \inf\{\|y - x\|: y \in \mathfrak{M}\}$. By Theorem 1 there exists a vector y_0 in $\mathfrak{M}$ such that $\|y_0 - x\| = \delta$; write $z = y_0 - x$. The fact that $z \neq 0$ follows from the fact that $x \epsilon' \mathfrak{M}$. Since $y_0 + \alpha y \in \mathfrak{M}$ for every vector y in $\mathfrak{M}$ and every complex number α, it follows that

$$\|z + \alpha y\| = \|(y_0 + \alpha y) - x\| \geqq \delta$$

and hence that

$$0 \leqq \|z + \alpha y\|^2 - \|z\|^2 = \alpha^*(z, y) + \alpha(y, z) + |\alpha|^2 \cdot \|y\|^2.$$

If, in particular, $\alpha = \beta(z, y)$ for any *real* number β, then

$$0 \leqq 2\beta\, |(y, z)|^2 + \beta^2\, |(y, z)|^2 \cdot \| y \|^2.$$

The validity of this inequality for small negative values of β implies the vanishing of the coefficient of the linear term. We conclude that $z \perp y$ and hence, since y is an arbitrary vector in $\mathfrak{M}$, that $z \perp \mathfrak{M}$.

§12. Orthogonal Complements

The *orthogonal complement* of a subset $\mathfrak{M}$ of $\mathfrak{H}$, in symbols $\mathfrak{M}^\perp$, is the set of all vectors x such that $x \perp \mathfrak{M}$. If $\mathfrak{M}$ and $\mathfrak{N}$ are subspaces such that $\mathfrak{M} \subset \mathfrak{N}$, the orthogonal complement of $\mathfrak{M}$ in $\mathfrak{N}$, in symbols $\mathfrak{N} - \mathfrak{M}$, is the set $\mathfrak{N} \cap \mathfrak{M}^\perp$.

THEOREM 1. *If $\mathfrak{M}$ is a subset of $\mathfrak{H}$, then $\mathfrak{M}^\perp$ is a subspace and $\mathfrak{M} \cap \mathfrak{M}^\perp \subset \mathfrak{O}$.*

Proof. If $x \in \mathfrak{M}$ and if y_1 and y_2 are in $\mathfrak{M}^\perp$, then, for every pair of complex numbers α_1 and α_2,

$$(x, \alpha_1 y_1 + \alpha_2 y_2) = \alpha_1^* (x, y_1) + \alpha_2^* (x, y_2) = 0,$$

so that $\mathfrak{M}^\perp$ is a linear manifold. The fact that $\mathfrak{M}^\perp$ is closed follows from the continuity of the inner product. To see that $\mathfrak{M} \cap \mathfrak{M}^\perp \subset \mathfrak{O}$, observe that if $x \in \mathfrak{M} \cap \mathfrak{M}^\perp$, then $x \perp x$.

THEOREM 2. *If $\mathfrak{M}$ is a subset of $\mathfrak{H}$, then $\mathfrak{M} \subset \mathfrak{M}^{\perp\perp}$.*

Proof. If $x \in \mathfrak{M}$ and $y \in \mathfrak{M}^\perp$, then $x \perp y$, so that $x \perp \mathfrak{M}^\perp$ and therefore $x \in \mathfrak{M}^{\perp\perp}$.

THEOREM 3. *If $\mathfrak{M}$ and $\mathfrak{N}$ are subsets of $\mathfrak{H}$ such that $\mathfrak{M} \subset \mathfrak{N}$, then $\mathfrak{M}^\perp \supset \mathfrak{N}^\perp$.*

THEOREM 4. *If $\mathfrak{M}$ is a subset of $\mathfrak{H}$, then $\mathfrak{M}^\perp = \mathfrak{M}^{\perp\perp\perp}$.*

Proof. Applying Theorem 2 to $\mathfrak{M}^\perp$ in place of $\mathfrak{M}$, we obtain $\mathfrak{M}^\perp \subset \mathfrak{M}^{\perp\perp\perp}$. Applying, on the other hand, Theorem 3 to the relation $\mathfrak{M} \subset \mathfrak{M}^{\perp\perp}$, we obtain the reverse inclusion $\mathfrak{M}^\perp \supset \mathfrak{M}^{\perp\perp\perp}$.

The preceding results are easy and in a sense automatic. As another such almost automatic result we mention the fact, whose proof is an easy exercise in the use of orthogonal complementation, that if $\{\mathfrak{M}_j\}$ is a family of subspaces, then $(\vee_j \mathfrak{M}_j)^\perp = \bigcap_j \mathfrak{M}_j^\perp$. The only non-trivial assertion along these lines (Theorem 5) is a consequence of the geometric discussion of the preceding section.

THEOREM 5. *If $\mathfrak{M}$ is a subspace, then $\mathfrak{M} = \mathfrak{M}^{\perp\perp}$.*

Proof. According to Theorem 2, $\mathfrak{M} \subset \mathfrak{M}^{\perp\perp}$. If $\mathfrak{M}$ were a proper subset of $\mathfrak{M}^{\perp\perp}$, then, by 11.2, $\mathfrak{M}^{\perp\perp}$ would have a non-zero vector in common with $\mathfrak{M}^{\perp}$; since this would contradict the relation $\mathfrak{M}^{\perp} \cap \mathfrak{M}^{\perp\perp} = \mathfrak{O}$, the proof is complete.

It is worth remarking that, applying the identity $(\bigvee_j \mathfrak{M}_j)^{\perp} = \bigcap_j \mathfrak{M}_j^{\perp}$ to the family $\{\mathfrak{M}_j^{\perp}\}$ in place of $\{\mathfrak{M}_j\}$, we obtain, in view of Theorem 5, the identity $(\bigcap_j \mathfrak{M}_j)^{\perp} = \bigvee_j \mathfrak{M}_j^{\perp}$.

To obtain the deepest and most useful fact about orthogonal complementation (Theorem 7), we need an auxiliary concept and an auxiliary result which are of considerable interest in themselves. The concept is that of the *vector sum* of two subspaces $\mathfrak{M}$ and $\mathfrak{N}$, in symbols $\mathfrak{M} + \mathfrak{N}$; it is defined to be the set of all vectors of the form $x + y$ with $x \in \mathfrak{M}$ and $y \in \mathfrak{N}$. It is easy to see that $\mathfrak{M} + \mathfrak{N} \subset \mathfrak{M} \vee \mathfrak{N}$ and that $\mathfrak{M} + \mathfrak{N}$ is a linear manifold; the result is that in at least one important case $\mathfrak{M} + \mathfrak{N}$ is actually a subspace. The hypothesis sufficient to guarantee this is that $\mathfrak{M}$ and $\mathfrak{N}$ are *orthogonal*, in symbols $\mathfrak{M} \perp \mathfrak{N}$: this means, naturally, that $x \perp \mathfrak{N}$ for every x in $\mathfrak{M}$.

THEOREM 6. *If $\mathfrak{M}$ and $\mathfrak{N}$ are orthogonal subspaces, then $\mathfrak{M} + \mathfrak{N}$ is closed.*

Proof. Suppose that $\{z_n\}$ is a sequence of vectors in $\mathfrak{M} + \mathfrak{N}$, so that, for each n, $z_n = x_n + y_n$ with $x_n \in \mathfrak{M}$ and $y_n \in \mathfrak{N}$, and suppose that the sequence $\{z_n\}$ converges to a vector z in $\mathfrak{H}$. By the Pythagorean theorem, $\| z_n - z_m \|^2 = \| x_n - x_m \|^2 + \| y_n - y_m \|^2$ for every n and m, and therefore both sequences $\{x_n\}$ and $\{y_n\}$ are Cauchy sequences. If $x_n \to x$ and $y_n \to y$, then $x \in \mathfrak{M}$ and $y \in \mathfrak{N}$; it follows from the continuity of addition that $z_n \to x + y$ and hence that $z \in \mathfrak{M} + \mathfrak{N}$.

THEOREM 7. (*The projection theorem.*) *If $\mathfrak{M}$ is a subspace, then $\mathfrak{M} + \mathfrak{M}^{\perp} = \mathfrak{H}$.*

Proof. If $\mathfrak{M} + \mathfrak{M}^{\perp} = \mathfrak{N}$, then, by Theorem 6, $\mathfrak{N}$ is a subspace. Since $\mathfrak{M} \subset \mathfrak{N}$ and $\mathfrak{M}^{\perp} \subset \mathfrak{N}$, it follows that $\mathfrak{N}^{\perp} \subset \mathfrak{M}^{\perp}$ and $\mathfrak{N}^{\perp} \subset \mathfrak{M}^{\perp\perp}$, and therefore that $\mathfrak{N}^{\perp} = \mathfrak{O}$. We conclude, as desired, that $\mathfrak{N} = \mathfrak{N}^{\perp\perp} = \mathfrak{O}^{\perp} = \mathfrak{H}$.

§13. Vector Sums

The concept of vector sums, introduced in the preceding section, deserves further study and generalization. The first step, namely the pertinent definition, is easy: we define the *vector sum* of an arbitrary family $\{\mathfrak{M}_j\}$ of subspaces, in symbols $\Sigma_j \mathfrak{M}_j$, to be the set of all vectors of the form $\Sigma_j x_j$ with $x_j \in \mathfrak{M}_j$ for all j. It is easy to see, just as in the finite case, that $\Sigma_j \mathfrak{M}_j$ is a linear manifold. As the following theorems show, the very close connection between vector sums and spans also persists in the general case.

THEOREM 1. *If* $\{\mathfrak{M}_j\}$ *is a family of subspaces and* $\mathfrak{M} = \Sigma_j \mathfrak{M}_j$, *then* $\bigvee_j \mathfrak{M}_j = \overline{\mathfrak{M}}$.

Proof. Since $\bigcup_j \mathfrak{M}_j \subset \overline{\mathfrak{M}}$, and since $\overline{\mathfrak{M}}$ is a subspace, it follows that $\bigvee_j \mathfrak{M}_j \subset \overline{\mathfrak{M}}$. Consider, on the other hand, the set of all those vectors of the form $\Sigma_j x_j$ for which $x_j \in \mathfrak{M}_j$ for all j and for which $x_j = 0$ for all but a finite number of values of j. Since, by the definition of infinite sums, this set of vectors is dense in $\mathfrak{M}$, and since $\bigvee_j \mathfrak{M}_j$ is closed, it follows that $\mathfrak{M} \subset \bigvee_j \mathfrak{M}_j$ and therefore that $\overline{\mathfrak{M}} \subset \bigvee_j \mathfrak{M}_j$.

We call attention to the fact that Theorem 1 is a non-trivial statement even in the finite case: it asserts that if $\mathfrak{M}_1$ and $\mathfrak{M}_2$ are subspaces and $\mathfrak{M} = \mathfrak{M}_1 + \mathfrak{M}_2$, then $\mathfrak{M}_1 \vee \mathfrak{M}_2 = \overline{\mathfrak{M}}$. We have seen that if $\mathfrak{M}_1$ and $\mathfrak{M}_2$ are orthogonal, then $\mathfrak{M}$ is closed and therefore $\mathfrak{M}_1 \vee \mathfrak{M}_2 = \mathfrak{M}$; it is natural to ask whether or not the bar (the closure operation) is ever really necessary. In §15 we shall show that it is, i.e. that the vector sum of two subspaces can fail to be a subspace.

We turn now to that part of the theory of vector sums which behaves itself—in which, that is, the pathology we mentioned in the preceding paragraph cannot occur. A family $\{\mathfrak{M}_j\}$ of subspaces is an *orthogonal family* if $\mathfrak{M}_j \perp \mathfrak{M}_k$ whenever $j \neq k$. (A vector sum of an orthogonal family of subspaces is frequently called an *orthogonal sum*, an *internal direct sum*, or simply a *direct sum*.)

THEOREM 2. *If* $\{\mathfrak{M}_j\}$ *is an orthogonal family of subspaces, then* $\bigvee_j \mathfrak{M}_j = \Sigma_j \mathfrak{M}_j$; *the representation of an element of* $\Sigma_j \mathfrak{M}_j$ *in the form* $\Sigma_j x_j$, *with* $x_j \in \mathfrak{M}_j$ *for all* j, *is unique.*

Proof. To prove the first part of the theorem, it is sufficient to show that $\bigvee_j \mathfrak{M}_j \subset \Sigma_j \mathfrak{M}_j$. If $x \in \bigvee_j \mathfrak{M}_j$, then, by the projection theorem, for each value of j there exists a vector x_j in $\mathfrak{M}_j$ and there exists a vector y_j in $\mathfrak{M}_j^{\perp}$ such that $x = x_j + y_j$. If $x_j \neq 0$ for some j, then $(x, x_j/\|x_j\|) = \|x_j\|$ and it follows therefore, from Bessel's inequality, that $\Sigma_j \|x_j\|^2 < \infty$. Applying 8.2, we see that there exists a vector x_0 such that $x_0 = \Sigma_j x_j$; we shall show that $x_0 = x$. If $y \in \mathfrak{M}_{j_0}$ for some j_0, then $(x - x_0, y) = (x_{j_0}, y) + (y_{j_0}, y) - (\Sigma_j x_j, y) = 0$ (by 7.3), i.e. $x - x_0 \perp \mathfrak{M}_{j_0}$ for all j_0. It follows that $x - x_0 \perp \Sigma_j \mathfrak{M}_j$ and therefore, by Theorem 1, that $x - x_0 \perp \bigvee_j \mathfrak{M}_j$. Since, however, $x - x_0 \in \bigvee_j \mathfrak{M}_j$, we conclude that indeed $x - x_0 = 0$. To prove the second part of the theorem, it is sufficient to show that if $\Sigma_j x_j = 0$, with $x_j \in \mathfrak{M}_j$ for all j, then $x_j = 0$ for all j, and this follows from the (general) Pythagorean theorem.

§14. Bases

A *basis* of a subspace $\mathfrak{M}$ is a maximal orthonormal family of vectors in $\mathfrak{M}$.

Although it follows immediately from Zorn's lemma that every subspace possesses a basis, it is sometimes possible to replace this transfinite argument by a constructive method; one such method is the *Gram-Schmidt orthogonalization process*. The process is an inductive one which, at its k-th stage, replaces the k-th term of a linearly independent sequence $\{x_n\}$ of vectors by a vector y_k in such a way that (i) y_k is a linear combination of $x_1, \cdots, x_k$, (ii) the sequence $\{y_n\}$ is orthonormal, and (iii) $\vee\{y_n\} = \vee\{x_n\}$. The process can be started off by writing $y_1 = x_1/\|x_1\|$; after $y_1, \cdots, y_k$ have been constructed, y_{k+1} is obtained by normalizing the vector $x_{k+1} - \Sigma_{j=1}^{k}(x_{k+1}, y_j)y_j$.

If μ is Lebesgue measure in the unit interval, and if

$$f_n(t) = t^n, \qquad 0 \leqq t \leqq 1, \quad n = 0, 1, 2, \cdots,$$

then the Gram-Schmidt orthogonalization process may be applied to the sequence $\{f_n\}$ in the Hilbert space $\mathfrak{L}_2(\mu)$. The process yields a basis of $\mathfrak{L}_2(\mu)$ consisting of polynomials. Another basis of $\mathfrak{L}_2(\mu)$ is the sequence $\{g_n\}$, where

$$g_n(t) = e^{2\pi int}, \qquad 0 \leqq t \leqq 1, \quad n = 0, \pm 1, \pm 2 \cdots.$$

In the Hilbert space of all families $\{\xi_j\}$ of complex numbers such that $\Sigma_j |\xi_j|^2 < \infty$, the vectors $\{\xi_j^{(k)}\}$ defined by $\xi_j^{(k)} = \delta_{jk}$ constitute a basis.

THEOREM 1. *A necessary and sufficient condition that an orthonormal family $\{x_j\}$ of vectors in a subspace $\mathfrak{M}$ satisfy all the following conditions is that it satisfy any one of them.*

(i) *The family $\{x_j\}$ is a basis of $\mathfrak{M}$.*

(ii) *If $x \in \mathfrak{M}$ and if $x \perp x_j$ for all j, then $x = 0$.*

(iii) *If, for each j, $\mathfrak{M}_j$ is the subspace spanned by the set consisting of the single vector x_j, then $\vee_j \mathfrak{M}_j = \mathfrak{M}$.*

(iv) *If $x \in \mathfrak{M}$, then $x = \Sigma_j(x, x_j)x_j$. (Fourier expansion.)*

(v) *If x and y are in $\mathfrak{M}$, then $(x, y) = \Sigma_j(x, x_j)(x_j, y)$. (Parseval's identity.)*

(vi) *If $x \in \mathfrak{M}$, then $\|x\|^2 = \Sigma_j |(x, x_j)|^2$.*

Proof. We shall prove that each of the conditions (i), (ii), (iii), (iv), and (v) implies the one following it and that (vi) implies (i).

(i) If $x \in \mathfrak{M}$, if $x \perp x_j$ for all j, and if $x \neq 0$, then $x/\| x \|$ may be adjoined to the family $\{x_j\}$, in contradiction to the assumed maximality of that family.

(ii) If $\vee_j \mathfrak{M}_j \neq \mathfrak{M}$, then, by 11.2, $\mathfrak{M}$ contains a non-zero vector x such that $x \perp x_j$ for all j.

(iii) Since $\{\mathfrak{M}_j\}$ is an orthogonal family of subspaces, 13.2 implies that $\vee_j \mathfrak{M}_j = \Sigma_j \mathfrak{M}_j$ and hence, if (iii) is true, every vector x in $\mathfrak{M}$ has the form $\Sigma_j \alpha_j x_j$ with suitable complex numbers α_j. It follows that $(x, x_k) = \Sigma_j \alpha_j (x_j, x_k) = \alpha_k$ for every index k.

(iv) If $x = \Sigma_j \alpha_j x_j$ and $y = \Sigma_j \beta_j x_j$, with $\alpha_j = (x, x_j)$ and $\beta_j = (y, x_j)$ for all j, then $(x, y) = (\Sigma_j \alpha_j x_j, \Sigma_k \beta_k x_k) = \Sigma_j \alpha_j \beta_j^*$.

(v) If (v) is true for all x and y in $\mathfrak{M}$, then it is true, in particular, when $x = y$.

(vi) If the family $\{x_j\}$ is not maximal, say, for instance, if it remains orthonormal after the adjunction of a vector x, then that vector x does not satisfy the relation (vi).

§15. A Non-closed Vector Sum

Familiarity with bases and Fourier expansions enables us to give an example of two subspaces $\mathfrak{M}$ and $\mathfrak{N}$ such that $\mathfrak{M} + \mathfrak{N} \neq \mathfrak{M} \vee \mathfrak{N}$.

To motivate the construction, we recall first of all that if $\mathfrak{M} \perp \mathfrak{N}$, then $\mathfrak{M} + \mathfrak{N} = \mathfrak{M} \vee \mathfrak{N}$, i.e. that equality holds if $\mathfrak{M}$ and $\mathfrak{N}$ are orthogonal. Equality can be made not to hold by getting as far as possible from orthogonality. Intuitively speaking we may say that the subspaces we shall construct make an angle of zero degrees; more precisely we shall construct $\mathfrak{M}$ and $\mathfrak{N}$ so that $\mathfrak{M} \cap \mathfrak{N} = \mathfrak{O}$ and, for suitable normalized vectors x and z, in $\mathfrak{M}$ and $\mathfrak{N}$ respectively, the inner product (x, z) comes arbitrarily near to 1.

Let $\{x_n\}$ and $\{y_m\}$ be two infinite orthonormal sequences such that $x_n \perp y_m$ for all n and m, and write $z_n = \alpha_n x_n + \beta_n y_n$ for every n, where the coefficients α_n and β_n will be determined presently. The first condition that we wish to put on α_n and β_n is that the sequence $\{z_n\}$, which is automatically orthogonal, shall be orthonormal as well, i.e. that $1 = \| z_n \|^2 = | \alpha_n |^2 + | \beta_n |^2$. For the sake of simplicity we shall insist that α_n and β_n shall be strictly positive real numbers. Since in that case $(x_n, z_n) = \alpha_n$ for every n, it follows that the subspaces $\mathfrak{M} = \vee \{x_n\}$ and $\mathfrak{N} = \vee \{z_n\}$ will certainly have the property mentioned in the preceding paragraph if $\alpha_n \to 1$, or equivalently $\beta_n \to 0$, as $n \to \infty$. For a technical reason (which will become apparent soon) we choose to ensure the validity of the relation $\beta_n \to 0$ by selecting the β_n's so that $\Sigma_n \beta_n^2 < \infty$. This is all the machinery we need; we remark that the sequences $\left\{\cos \frac{1}{n}\right\}$ and $\left\{\sin \frac{1}{n}\right\}$, for example, have all the properties we demand of $\{\alpha_n\}$ and $\{\beta_n\}$ respectively.

To prove that $\mathfrak{M} + \mathfrak{N} \neq \mathfrak{M} \vee \mathfrak{N}$, we have to exhibit a vector y in $\mathfrak{M} \vee \mathfrak{N}$ such that y does not belong to $\mathfrak{M} + \mathfrak{N}$. Since $\Sigma_m \beta_m^2 < \infty$, it follows that the sequence $\{\beta_m y_m\}$ is summable; we assert that if $y = \Sigma_m \beta_m y_m$, then the vector y has the desired properties. The fact that $\alpha_m \neq 0$ implies, indeed, that $y_m \in \mathfrak{M} + \mathfrak{N}$ for every m and hence that $y \in \mathfrak{M} \vee \mathfrak{N}$. If it were true, however, that $y \in \mathfrak{M} + \mathfrak{N}$, say $y = x + z$ with $x \in \mathfrak{M}$ and $z \in \mathfrak{N}$, then we should have

$$\beta_m = (y, y_m) = (x + z, y_m) = (z, y_m) = (\Sigma_n(z, z_n)z_n, y_m)$$
$$= (z, z_m)(z_m, y_m) = (z, z_m)\beta_m$$

for every m. Since $\beta_m \neq 0$, it would then follow that $(z, z_m) = 1$ for every m, but since (z, z_m) is the m-th Fourier coefficient of z with respect to $\{z_m\}$, this is preposterous.

§16. Dimension

THEOREM 1. *Any two bases of a subspace $\mathfrak{M}$ have the same power.*

Proof. Let $\{x_j\}$ and $\{y_k\}$ be two bases of $\mathfrak{M}$, of powers u and v respectively. Since $x_j = \Sigma_k(x_j, y_k)y_k$ for each j, the set K_j of those indices k for which $(x_j, y_k) \neq 0$ is countable. Since $y_k \in \mathfrak{M} = \bigvee \{x_j\}$, no y_k can be orthogonal to all x_j, i.e. every index k is contained in $\bigcup_j K_j$. It follows that $v \leqq \aleph_0 \cdot u$ and, by symmetry, $u \leqq \aleph_0 \cdot v$. If both u and v are infinite, the proof is complete; if either u or v is finite, the theorem reduces to a known result in the theory of finite-dimensional vector spaces.

Theorem 1 allows us to define the dimension of a subspace $\mathfrak{M}$ as the common power of all bases of $\mathfrak{M}$. In the remainder of this section we propose to show (Theorem 3) that in a sense the dimension of the Hilbert space $\mathfrak{H}$ completely determines the structure of $\mathfrak{H}$.

An *isomorphism* from a Hilbert space $\mathfrak{H}$ onto a Hilbert space $\mathfrak{K}$ is a one-to-one linear transformation U from $\mathfrak{H}$ onto $\mathfrak{K}$ such that $(Ux, Uy) =$

(x, y) for every pair of vectors x and y in $\mathfrak{H}$; an *isometry* from a Hilbert space $\mathfrak{H}$ to a Hilbert space $\mathfrak{K}$ is a linear transformation U from $\mathfrak{H}$ into $\mathfrak{K}$ such that $\| Ux \| = \| x \|$ for every vector x in $\mathfrak{H}$. Observe that an isometry deserves its name, i.e. that, in virtue of the equation $\| Ux - Uy \| = \| U(x - y)\| = \| x - y \|$, an isometry preserves not only norms (distances from 0) but all distances. Observe also that an isomorphism is necessarily an isometry. Since an isometry from $\mathfrak{H}$ to $\mathfrak{K}$ need not map $\mathfrak{H}$ onto $\mathfrak{K}$, it is easy to construct isometries which are not isomorphisms; our next result shows that the into-onto distinction is the only one between isomorphisms and isometries.

THEOREM 2. *A linear transformation U from a Hilbert space $\mathfrak{H}$ to a Hilbert space $\mathfrak{K}$ is an isomorphism if and only if it is an isometry, mapping $\mathfrak{H}$ onto $\mathfrak{K}$.*

Proof. We have already seen that an isomorphism is an isometry. If, conversely, U is an isometry, and if $Ux = Uy$, then $0 = \| U(x - y)\| = \| x - y \|$, and it follows therefore that U is one-to-one. The fact that U preserves inner products follows from the assumption that if $\varphi(x, y) = (Ux, Uy)$ and $\psi(x, y) = (x, y)$, then the bilinear functionals φ and ψ induce the same quadratic form.

Two Hilbert spaces are called *isomorphic* if there exists an isomorphism between them. It follows from the definition of an isomorphism and from our observations concerning isomorphisms and isometries that an isomorphism preserves all the structure that went into the definition of Hilbert spaces and that, consequently, isomorphic Hilbert spaces are geometrically indistinguishable and may legitimately be viewed as identical.

THEOREM 3. *Two Hilbert spaces are isomorphic if and only if they have the same dimension.*

Proof. In view of the intrinsic definition of dimension, the "only if" part is obvious. Suppose, conversely, that $\mathfrak{H}$ and $\mathfrak{K}$ are Hilbert spaces of the same dimension and let $\{x_j\}$ and $\{y_j\}$ be bases of $\mathfrak{H}$ and $\mathfrak{K}$ respectively, indexed by the same set $\{j\}$. If $x = \Sigma_j \alpha_j x_j$ is any vector in $\mathfrak{H}$ and if Ux is defined to be $\Sigma_j \alpha_j y_j$, then U is clearly a linear transformation from $\mathfrak{H}$ onto $\mathfrak{K}$; since $\| Ux \|^2 = \Sigma_j | \alpha_j |^2 = \| x \|^2$, U is an isometry. The proof is completed by an application of Theorem 2.

According to Theorem 3 any property that some Hilbert spaces do and others do not possess can be characterized simply by counting. Thus, for instance, a necessary and sufficient condition that $\mathfrak{H}$ be separable is that the dimension of $\mathfrak{H}$ be not greater than $\aleph_0$. Indeed, since the distance between any two terms of an orthonormal family is $\sqrt{2}$, it follows that if $\mathfrak{H}$ is separable, then no orthonormal family can be uncountable. If, on the other hand, a countable, maximal orthonormal family $\{x_j\}$ exists, then the set of all finite linear combinations, with coefficients whose real and imaginary parts are both rational, is a countable dense set in $\mathfrak{H}$.

§17. Boundedness

A linear transformation A from a Hilbert space $\mathfrak{H}$ to a Hilbert space $\mathfrak{K}$ is *bounded* if there exists a positive real number α such that $\| Ax \| \leqq \alpha \| x \|$ for all x in $\mathfrak{H}$; the *norm* of A, in symbols $\| A \|$, is the infimum of all such values of α.

THEOREM 1. *A linear transformation A from a Hilbert space $\mathfrak{H}$ to a Hilbert space $\mathfrak{K}$ is bounded if and only if it maps the* **unit sphere** *(i.e. the set $\{x: \| x \| = 1\}$) onto a bounded subset of $\mathfrak{K}$; if $\alpha = \sup \{\| Ax \| : \| x \| = 1\}$, then $\| A \| = \alpha$.*

Proof. If A is bounded and $\| x \| = 1$, then $\| Ax \| \leqq \| A \| \cdot \| x \| = \| A \|$ and therefore $\alpha \leqq \| A \|$. If, conversely, $\alpha < \infty$, then, for every non-zero vector x,

$$\| Ax \| = \| A(\| x \| \cdot (x / \| x \|)) \| = \| A(x / \| x \|) \| \cdot \| x \| \leqq \alpha \| x \|,$$

so that A is bounded and $\| A \| \leqq \alpha$.

THEOREM 2. *A linear transformation A from a Hilbert space $\mathfrak{H}$ to a Hilbert space $\mathfrak{K}$ is bounded if and only if it is continuous.*

Proof. If A is bounded, then its continuity follows from the relation $\| Ax - Ay \| \leqq \| A \| \cdot \| x - y \|$, valid for all vectors x and y in $\mathfrak{H}$. If A is not bounded, then, for every positive integer n, there exists a vector x_n in $\mathfrak{H}$ such that $\| x_n \| = 1$ and $\| Ax_n \| \geqq n$. Since $\frac{1}{n} x_n \to 0$, whereas $\left\| A \left(\frac{1}{n} x_n \right) \right\| \geqq 1$, it follows that A is not continuous at 0.

The definition of boundedness and Theorems 1 and 2 apply in particular to linear transformations from a Hilbert space $\mathfrak{H}$ to the special one-dimensional Hilbert space $\mathfrak{C}$, i.e. to linear functionals. In this special case there is available to us a powerful and elegant result which completely characterizes all bounded linear functionals.

THEOREM 3. (*The Riesz representation theorem for bounded linear functionals.) A linear functional* ξ *on* $\mathfrak{H}$ *is bounded if and only if there exists a vector* y *such that* $\xi(x) = (x, y)$ *for all* x*; such a* y*, if it exists, is unique.*

Proof. If $\xi(x) = (x, y)$ for all x, then $|\xi(x)| \leqq \|x\| \cdot \|y\|$, so that ξ is bounded and, in fact, $\|\xi\| \leqq \|y\|$. (It is easy, but for our purposes unnecessary, to prove that $\|\xi\| = \|y\|$.) The uniqueness of y follows from 4.1.

If, conversely, ξ is a bounded linear functional and if $\mathfrak{M} = \{x : \xi(x) = 0\}$, then $\mathfrak{M}$ is a subspace. If $\mathfrak{M} = \mathfrak{H}$, then $\xi(x)$ is indeed identically equal to (x, y) with $y = 0$. If $\mathfrak{M} \neq \mathfrak{H}$, then $\mathfrak{M}^{\perp}$ contains a non-zero vector z; we shall prove that a suitable multiple αz of z is an admissible y. No matter what the value of α is, it is clear that if $y = \alpha z$, then $\xi(x) = 0 = (x, y)$ whenever $x \in \mathfrak{M}$. If, on the other hand, $x = \beta z$ for some complex number β, then $(x, y) = (\beta z, \alpha z) = \alpha^* \beta \|z\|^2$, so that a necessary and sufficient condition for the validity of the identity $\xi(\beta z) = (\beta z, y)$ is that $\alpha = \xi^*(z)/\|z\|^2$. With this choice of α it is then true that $\xi(x) = (x, y)$, with $y = \alpha z$, if either $x \in \mathfrak{M}$ or x is a multiple of z. Since for an arbitrary vector x in $\mathfrak{H}$, $x - \beta z \in \mathfrak{M}$ if $\beta = \xi(x)/\xi(z)$ (note that $\xi(z) \neq 0$), it follows that $\xi(x) = \xi(x - \beta z) + \xi(\beta z) = (x - \beta z, y) + (\beta z, y) = (x, y)$.

§18. Bounded Bilinear Functionals

Since a linear functional is a linear transformation, any meaningful statement that applies to all linear transformations applies, in particular, to linear functionals. Since bilinear functionals and quadratic forms are *not* linear transformations, their theory is not a special case but merely an analog of the theory of linear transformations. The analogy is quite close. We shall, for instance, say that a bilinear functional φ is *bounded* if there exists a positive real number α such that $|\varphi(x, y)| \leqq \alpha \|x\| \cdot \|y\|$ for every pair of vectors x and y in $\mathfrak{H}$, and we define the *norm* of φ, in symbols $\|\varphi\|$, as the infimum of all such values of α. We shall also say that a quadratic form $\hat{\varphi}$ is *bounded* if there exists a positive real number α such that $|\hat{\varphi}(x)| \leqq \alpha \|x\|^2$ for all x in $\mathfrak{H}$; the *norm* of $\hat{\varphi}$, in symbols $\|\hat{\varphi}\|$, is the infimum of all such values of α. The first result of the preceding section may be stated (and proved) in almost exactly the same way for bilinear functionals and quadratic forms as for linear transformations.

THEOREM 1. *If* φ *is a bilinear functional on* $\mathfrak{H}$ *and if*

$$\alpha = \sup \{|\varphi(x, y)| : \|x\| = \|y\| = 1\},$$

then a necessary and sufficient condition that φ *be bounded is that* $\alpha < \infty$; *if* φ *is bounded, then* $\|\varphi\| = \alpha$. *If* $\hat{\varphi}$ *is a quadratic form on* $\mathfrak{H}$ *and if*

$$\alpha = \sup\{|\hat{\varphi}(x)| : \|x\| = 1\},$$

then a necessary and sufficient condition that $\hat{\varphi}$ *be bounded is that* $\alpha < \infty$; *if* $\hat{\varphi}$ *is bounded, then* $\|\hat{\varphi}\| = \alpha$.

The interesting and useful results along these lines concern the relations between the norm of a bilinear functional and the norm of its induced quadratic form.

THEOREM 2. *The quadratic form* $\hat{\varphi}$ *induced by a bilinear functional* φ *is bounded if and only if* φ *is bounded; if* φ *and* $\hat{\varphi}$ *are bounded, then*

$$\|\hat{\varphi}\| \leqq \|\varphi\| \leqq 2\|\hat{\varphi}\|.$$

Proof. If φ is bounded, then $|\hat{\varphi}(x)| = |\varphi(x, x)| \leqq \|\varphi\| \cdot \|x\| \cdot \|x\|$ for all x; it follows that $\hat{\varphi}$ is bounded and that $\|\hat{\varphi}\| \leqq \|\varphi\|$. If, conversely, $\hat{\varphi}$ is bounded, then, by polarization,

$$|\varphi(x, y)| \leqq \tfrac{1}{4}\|\hat{\varphi}\| \cdot (\|x + y\|^2 + \|x - y\|^2 + \|x + iy\|^2 + \|x - iy\|^2)$$

and hence, by the parallelogram law,

$$|\varphi(x, y)| \leqq \|\hat{\varphi}\| \cdot (\|x\|^2 + \|y\|^2)$$

for every pair of vectors x and y. It follows that $|\varphi(x, y)| \leqq 2\|\hat{\varphi}\|$ whenever $\|x\| = \|y\| = 1$, and consequently (by Theorem 1) φ is bounded and $\|\varphi\| \leqq 2\|\hat{\varphi}\|$.

It is not difficult to construct examples (in finite-dimensional spaces) to show that the inequalities in Theorem 2 are in general best possible. They allow, however, a considerable improvement in the symmetric case.

THEOREM 3. *If* φ *is a bounded, symmetric, bilinear functional, then* $\|\varphi\| = \|\hat{\varphi}\|$.

Proof. We need only prove that $\|\varphi\| \leqq \|\hat{\varphi}\|$. Since the symmetry of φ implies that $\hat{\varphi}$ is real, polarization shows that the real part of φ is given by the equation

$$\Re\varphi(x, y) = \hat{\varphi}(\tfrac{1}{2}(x + y)) - \hat{\varphi}(\tfrac{1}{2}(x - y)).$$

It follows that

$$|\Re\varphi(x, y)| \leqq \tfrac{1}{4}\|\hat{\varphi}\| \cdot (\|x + y\|^2 + \|x - y\|^2) = \tfrac{1}{2}\|\hat{\varphi}\| \cdot (\|x\|^2 + \|y\|^2),$$

and therefore that $|\Re\varphi(x, y)| \leqq \|\hat{\varphi}\|$ whenever $\|x\| = \|y\| = 1$. For an arbitrary, but temporarily fixed, pair of vectors x and y with $\|x\| = \|y\| = 1$, let θ be a complex number of absolute value 1 such that $\theta\varphi(x, y) = |\varphi(x, y)|$. The inequality just derived, when applied to θx and y, implies that

$$|\varphi(x, y)| = \varphi(\theta x, y) = |\Re\varphi(\theta x, y)| \leqq \|\hat{\varphi}\|,$$

and therefore the proof may be completed by an application of Theorem 1.

CHAPTER II

THE ALGEBRA OF OPERATORS

§19. Operators

An *operator* is a bounded linear transformation from $\mathfrak{H}$ into $\mathfrak{H}$.

THEOREM 1. *If A and B are operators and if, for every vector x and for every complex number α, $(\alpha A)x = \alpha(Ax)$, $(A + B)x = Ax + Bx$, and $(AB)x = A(Bx)$, then αA, $A + B$, and AB are operators such that $\| \alpha A \| = | \alpha | \cdot \| A \|$, $\| A + B \| \leqq \| A \| + \| B \|$, and $\| AB \| \leqq \| A \| \cdot \| B \|$.*

Proof. It is obvious that αA, $A + B$, and AB are linear transformations from $\mathfrak{H}$ into $\mathfrak{H}$. The fact that they are bounded, and that their norms behave as asserted, follows from the relations

$$\| \alpha(Ax) \| = | \alpha | \cdot \| Ax \|, \| Ax + Bx \| \leqq \| Ax \| + \| Bx \| \leqq (\| A \| + \| B \|) \cdot \| x \|,$$

and

$$\| A(Bx) \| \leqq \| A \| \cdot \| Bx \| \leqq \| A \| \cdot \| B \| \cdot \| x \|.$$

A painless verification shows that the set of all operators on $\mathfrak{H}$ is a complex vector space with respect to the scalar multiplication and addition defined in Theorem 1, and that the multiplication there described is associative and bilinear—in other words that with respect to these operations the set of all operators on $\mathfrak{H}$ is an algebra. This algebra contains a unit, called the *identity* operator and denoted by the symbol 1; it is defined by writing $1x = x$ for all x. No confusion will arise from using the same symbol for an operator as for a number, nor even from generalizing this notation and, for any complex number α, using the symbol α to denote also the operator $\alpha 1$. Observe that we are thereby committed to using the symbol 0 for the operator such that $0x = 0$ for all x.

As in every algebra, we shall use the symbol A^n to denote the product of n factors all equal to A, $n = 1, 2, \cdots$; A^0 is defined to be 1. More generally if p is any (complex) polynomial, $p(\lambda) = \Sigma_{j=0}^{n} \alpha_j \lambda^j$, we shall use the symbol $p(A)$ for the operator $\Sigma_{j=0}^{n} \alpha_j A^j$.

To these algebraic remarks we adjoin a result concerning a useful aspect of the most important topological property (i.e. continuity) that an operator possesses.

THEOREM 2. *If A is an operator, x is a vector, and $\{x_j\}$ is a family of vectors such that $\Sigma_j x_j = x$, then $\Sigma_j A x_j = Ax$.*

Proof. For any positive number ε we may find a finite set J_0 of indices such that $\| x - \Sigma_{j\epsilon J} x_j \| < \varepsilon$ whenever J is a finite set of indices containing J_0. It follows that $\| Ax - \Sigma_{j\epsilon J} Ax_j \| \leqq \| A \|\varepsilon$ whenever $J \supset J_0$, and this implies that $\Sigma_j Ax_j = Ax$.

§20. Examples of Operators

Since considerations concerning operators will occupy us during most of the remainder of this book, it might be a good idea to look at a few of them.

(i) One of the most classical examples is obtained as follows. Let X be a measure space with measure μ, and let h be a complex-valued measurable function on the Cartesian product of X with itself, square-integrable with respect to the product measure in that Cartesian product space. If $f \epsilon \mathfrak{L}_2(\mu)$, and if $Af = g$, where $g(s) = \int h(s, t)f(t)\, d\mu(t)$, then A is an operator on $\mathfrak{L}_2(\mu)$.

(ii) Another operator on the space $\mathfrak{L}_2(\mu)$ is obtained by selecting a fixed, essentially bounded, measurable function h on X and writing $Af = g$, where $g(t) = h(t)f(t)$. Operators of this type are of sufficiently general interest and importance to deserve a name; we shall refer to the operator A as the *multiplication operator*, or simply the *multiplication*, defined by h.

If X is the interior of the unit circle in the complex plane, if $h(\lambda) = \lambda$ for every λ in X, if μ is Lebesgue measure, and if instead of $\mathfrak{L}_2(\mu)$ we consider the subspace of analytic functions described in §9(i), we obtain an interesting and significantly different variant of this example.

(iii) For another example, let T be a one-to-one measure-preserving transformation of X onto itself and write $Af = g$, where $g(s) = f(Ts)$. To obtain an easily manageable special case, let X be the real line, let μ be Lebesgue measure, and define T by $Ts = s + 1$. A useful generalization of this special case is obtained by replacing the real line by any locally compact topological group, replacing μ by its left Haar measure, and defining T to be, say, left multiplication by a fixed element.

(iv) Consider the Hilbert space of all sequences $\{\xi_n\}$ of complex numbers ($n = 1, 2, \cdots$) such that $\Sigma_n |\xi_n|^2 < \infty$, and write $A\{\xi_n\} = \{\eta_n\}$, where $\eta_n = \xi_{n+1}$ for all n. Formally the same definition of A yields a significantly different operator if we consider instead the Hilbert space of all families $\{\xi_n\}$ of complex numbers, $n = 0, \pm 1, \pm 2, \cdots$. Instead of writing $\eta_n = \xi_{n+1}$, we might have written $\eta_n = \Sigma_m \alpha_{nm} \xi_m$, where $\alpha_{nm} = \delta_{n+1,m}$; different examples of operators are obtained by varying the matrix $[\alpha_{nm}]$. We shall not enter into a discussion of what conditions a matrix must satisfy in order to define an operator, but, by way of a hint that will at least yield a sufficient condition, we remark that the operators defined by matrices are special cases of operators defined by integral kernels; cf. example (i).

§21. Inverses

An operator A is *invertible* if there exists an operator B such that $AB = BA = 1$. The reader should be as competent as the author at constructing examples of operators which are and of operators which are not invertible simply by examining the examples given in the preceding section.

THEOREM 1. *If A, B, and C are operators such that $AB = CA = 1$, then $B = C$, and consequently A is invertible.*

Proof. $B = 1 \cdot B = (CA)B = C(AB) = C \cdot 1 = C$.

It follows from Theorem 1 that if an operator A is invertible, then there exists only one operator B such that $BA = AB = 1$; we shall write $B = A^{-1}$ and call A^{-1} the *inverse* of A. Standard elementary considerations prove that if A and B are invertible operators and if n is a positive integer, then the operators A^{-1}, AB, and A^n are invertible, and their inverses are given by the equations $(A^{-1})^{-1} = A$, $(AB)^{-1} = B^{-1}A^{-1}$, and $(A^n)^{-1} = (A^{-1})^n$. In view of the last relation we may consistently define A^n, for invertible operators A and negative integers n, by $A^n = (A^{-1})^{-n}$.

It is useful to have at hand some geometric conditions for invertibility; such conditions can be given in terms of the range of an operator. Recall that the *range* of an operator A is the set of all vectors of the form Ax; the range of an operator is always a linear manifold, but it is not necessarily a subspace.

THEOREM 2. *If A is an operator and α is a positive real number such that $\|Ax\| \geqq \alpha \|x\|$ for every vector x, then the range of A is closed.*

Proof. If $y_n = Ax_n$, $n = 1, 2, \cdots$, and if $y_n \to y$, then, since we have $\| y_n - y_m \| = \| Ax_n - Ax_m \| \geqq \alpha \| x_n - x_m \|$ for all n and m, it follows that $\{x_n\}$ is a Cauchy sequence and hence that there exists a vector x such that $x_n \to x$. The continuity of A implies that $y = Ax$ and hence that y is in the range of A.

THEOREM 3. *An operator A is invertible if and only if its range is dense in $\mathfrak{H}$ and there exists a positive real number α such that $\| Ax \| \geqq \alpha \| x \|$ for every vector x.*

Proof. If A is invertible and if $y \in \mathfrak{H}$, write $x = A^{-1}y$; since $Ax = y$, it follows that the range of A is not only dense in $\mathfrak{H}$ but, in fact, coincides with $\mathfrak{H}$. It follows also that, for every vector x,

$$\| x \| = \| A^{-1}Ax \| \leqq \| A^{-1} \| \cdot \| Ax \|,$$

i.e. that the condition of the theorem is satisfied with $\alpha = 1/\| A^{-1} \|$. Suppose now that the range of A is dense and that $\| Ax \| \geqq \alpha \| x \|$ for all x. According to Theorem 2 we may conclude that the range of A is in fact equal to $\mathfrak{H}$. If $Ax_1 = Ax_2$, i.e. $Ax_1 - Ax_2 = 0$, then

$$0 = \| Ax_1 - Ax_2 \| \geqq \alpha \| x_1 - x_2 \|,$$

and therefore $x_1 = x_2$. This implies that not only is it true that every vector y in $\mathfrak{H}$ has the form Ax for some x in $\mathfrak{H}$, but in fact there is exactly one such x, and a single-valued transformation B of $\mathfrak{H}$ into itself is defined by writing $By = x$. Since B is easily verified to be linear, and since $\| y \| = \| Ax \| \geqq \alpha \| x \| = \alpha \| By \|$, it follows that B is an operator (and we even obtain the inequality $\| B \| \leqq 1/\alpha$). The relations $ABy = Ax = y$ and $BAx = By = x$ show that $AB = BA = 1$, and hence that A is invertible (and we even obtain the result $B = A^{-1}$).

§22. Adjoints

If A is a (not necessarily bounded) linear transformation from $\mathfrak{H}$ into $\mathfrak{H}$, and if $\varphi(x, y) = (Ax, y)$ for every pair of vectors x and y, then φ is a bilinear functional. The elementary properties of the inner product imply that if A_1 and A_2 are two linear transformations from $\mathfrak{H}$ into $\mathfrak{H}$ such that $(A_1x, y) = (A_2x, y)$ for all x and y, then $A_1 = A_2$. These facts together with 3.2 show that if only $(A_1x, x) = (A_2x, x)$ for all x, then already $A_1 = A_2$. We begin the proper business of this section by showing that the connection between linear transformations and bilinear functionals goes quite a bit deeper than these superficial remarks.

THEOREM 1. *If A is an operator and if $\varphi(x, y) = (Ax, y)$ for all x*

and y, *then* φ *is a bounded bilinear functional and* $\|\varphi\| = \|A\|$. *If, conversely,* φ *is a bounded bilinear functional, then there exists a unique operator* A *such that* $\varphi(x, y) = (Ax, y)$ *for all* x *and* y.

Proof. If A is an operator and if $\varphi(x, y) = (Ax, y)$, then $|\varphi(x, y)| \leqq \|A\| \cdot \|x\| \cdot \|y\|$ for all x and y and consequently $\|\varphi\| \leqq \|A\|$. If, conversely, φ is a bounded bilinear functional and if $\eta_x(y) = \varphi(x, y)$ for all x and y, then, for each fixed x, $\eta_x{}^*$ is a bounded linear functional. It follows from the Riesz representation theorem (17.3) that there exists a unique vector Ax such that $\varphi(x, y) = (Ax, y)$ for all y. The linearity of the transformation A thereby defined is easily verified; its uniqueness follows from our remarks at the beginning of this section. Since

$$\|Ax\|^2 = (Ax, Ax) = \varphi(x, Ax) \leqq \|\varphi\| \cdot \|x\| \cdot \|Ax\|,$$

it follows that $\|Ax\| \leqq \|\varphi\| \cdot \|x\|$ for all x. But this implies that A is bounded and $\|A\| \leqq \|\varphi\|$, so that the proof is complete.

Observe that it follows from the first part of Theorem 1, together with 18.1, that $\|A\| = \sup\{|(Ax, y)| : \|x\| = \|y\| = 1\}$ for any operator A.

THEOREM 2. *If* A *is an operator, then there exists a unique operator* A^*, *called the* ***adjoint*** *of* A, *such that* $(Ax, y) = (x, A^*y)$ *for all* x *and* y; A^* *is such that* $\|A^*\| = \|A\|$.

Proof. Write $\varphi(x, y) = (Ax, y)$ and $\psi(x, y) = \varphi^*(y, x)$ for all x and y. Since, by Theorem 1, φ is a bounded bilinear functional, and since this implies that ψ is a bounded bilinear functional with $\|\psi\| = \|\varphi\| = \|A\|$, it follows from the converse part of Theorem 1 that there exists an operator A^* such that $\psi(x, y) = (A^*x, y)$ for all x and y and that A^* is such that $\|A^*\| = \|\psi\| = \|A\|$. Since the uniqueness of A^* is clear, the proof is completed by the obvious computation: $(Ax, y) = \varphi(x, y) = \psi^*(y, x) = (A^*y, x)^* = (x, A^*y)$.

The behavior of adjoints can be understood by constructing the adjoints of the various operators described in §20. We call special attention to the example of a multiplication restricted to the analytic functions: its adjoint is not what at first it might appear to be.

THEOREM 3. *If* A *and* B *are operators and* α *is a complex number, then*

(i) $A^{**} = A$,
(ii) $(\alpha A)^* = \alpha^* A^*$,
(iii) $(A + B)^* = A^* + B^*$,

and

(iv) $(AB)^* = B^*A^*$;

(v) *if* A *is invertible, then* A^* *is invertible and* $(A^*)^{-1} = (A^{-1})^*$.

Proof. Each of the five assertions is implied by the corresponding one of the following five identities.

(i) $(A^*x, y) = (y, A^*x)^* = (Ay, x)^* = (x, Ay)$.

(ii) $(\alpha Ax, y) = \alpha(Ax, y) = \alpha(x, A^*y) = (x, \alpha^*A^*y)$.

(iii) $((A + B)x, y) = (Ax, y) + (Bx, y) = (x, A^*y) + (x, B^*y) = (x, A^*y + B^*y) = (x, (A^* + B^*)y)$.

(iv) $(ABx, y) = (Bx, A^*y) = (x, B^*A^*y)$.

(v) $(A^{-1})^*A^* = (AA^{-1})^*$ and $A^*(A^{-1})^* = (A^{-1}A)^*$.

THEOREM 4. *If* A *is an operator, then* $\| A^*A \| = \| A \|^2$.

Proof. It follows from Theorem 2 that $\| A^*A \| \leqq \| A^* \| \cdot \| A \| = \| A \|^2$. On the other hand $\| Ax \|^2 = (Ax, Ax) = (A^*Ax, x) \leqq \| A^*A \| \cdot \| x \|^2$ for every vector x and therefore $\| A \|^2 \leqq \| A^*A \|$.

§23. Invariance

A subspace $\mathfrak{M}$ is *invariant* under an operator A if $A\mathfrak{M} \subset \mathfrak{M}$, i.e. if $Ax \in \mathfrak{M}$ whenever $x \in \mathfrak{M}$; a subspace $\mathfrak{M}$ *reduces* an operator A if both $\mathfrak{M}$ and $\mathfrak{M}^\perp$ are invariant under A.

THEOREM 1. *If each of a family* $\{\mathfrak{M}_j\}$ *of subspaces is invariant under an operator* A [*or reduces* A], *then* $\vee_j \mathfrak{M}_j$ *and* $\cap_j \mathfrak{M}_j$ *are both invariant under* A [*or reduce* A].

THEOREM 2. *A necessary and sufficient condition that a subspace* $\mathfrak{M}$ *be invariant under an operator* A *is that* $\mathfrak{M}^\perp$ *be invariant under* A^*.

Proof. By symmetry it is sufficient to prove that the condition is necessary. If $\mathfrak{M}$ is invariant under A, and if $x \in \mathfrak{M}$ and $y \in \mathfrak{M}^\perp$, then $(x, A^*y) = (Ax, y) = 0$, so that $A^*y \in \mathfrak{M}^\perp$, and consequently $\mathfrak{M}^\perp$ is invariant under A^*.

We record for later reference an immediate corollary of Theorem 2.

THEOREM 3. *A necessary and sufficient condition that a subspace* $\mathfrak{M}$ *reduce an operator* A *is that it be invariant under both* A *and* A^*.

The difference between invariance and reduction is somewhat subtle and it is worth while to take a close look at an example. Consider a Hilbert space which has an infinite sequence $\{x_n\}$ as a basis, $n = 1, 2, \cdots$, and define an operator A by $A(\Sigma_n \alpha_n x_n) = \Sigma_n \alpha_n x_{n+1}$; cf. §20 (iv). There are many non-trivial subspaces invariant under A; concrete examples may be obtained by selecting a fixed positive integer m and forming the subspace $\bigvee\{x_{n+m}: n = 1, 2, \cdots\}$. We assert, however, that the only subspaces which reduce A are $\mathfrak{O}$ and $\mathfrak{H}$, or, in other words, that if a subspace $\mathfrak{M}$ reduces A and contains a non-zero vector x, then $\mathfrak{M} = \mathfrak{H}$. To prove this it is convenient to employ Theorem 3 and it is necessary, therefore, to discover A^*; an immediate computation shows that A^* is defined by $A^*(\Sigma_n \alpha_n x_n) = \Sigma_n \alpha_{n+1} x_n$. If $\Sigma_n \alpha_n x_n$ is the Fourier expansion of the given non-zero vector x and if m is the lowest positive integer such that $\alpha_m \neq 0$, then $x = \Sigma_n \alpha_n x_n = \Sigma_n \alpha_{n+m-1} x_{n+m-1}$. Since the assumption that $\mathfrak{M}$ reduces A, together with Theorem 3, shows that $y = \Sigma_n \alpha_{n+m} x_{n+m} = A^m (A^*)^m x \in \mathfrak{M}$, it follows that $\alpha_m x_m = x - y \in \mathfrak{M}$, and hence that $x_m \in \mathfrak{M}$. Another application of the same reasoning shows that $x_n = A^{n-1}(A^*)^{m-1} x_m \in \mathfrak{M}$ for all n and it follows indeed that $\mathfrak{M} = \mathfrak{H}$.

§24. Hermitian Operators

An operator A is *Hermitian if* $A = A^*$.

THEOREM 1. *A necessary and sufficient condition that an operator A be Hermitian is that the bilinear functional φ, defined for every pair of vectors x and y by $\varphi(x, y) = (Ax, y)$, be symmetric.*

Proof. A necessary and sufficient condition that $\varphi(x, y) = \varphi^*(y, x)$ for all x and y is that $(Ax, y) = (y, A^*x)^* = (A^*x, y)$ for all x and y.

As an immediate consequence of Theorem 1 and what we already know about bilinear functionals (cf. 3.3, 18.3, and 22.1) we obtain the following characterization of Hermitian operators and their norms.

THEOREM 2. *An operator A is Hermitian if and only if (Ax, x) is real for every vector x; if A is Hermitian, then $\| A \| = \sup \{ | (Ax, x) | : \| x \| = 1 \}$.*

Most of the algebraic properties of the set of Hermitian operators follow quite trivially from the definition. It is, for instance, clear that a real scalar multiple of a Hermitian operator and the sum of two Hermitian operators are Hermitian, and that the inverse of an invertible Hermitian operator is also Hermitian. To describe the situation concerning products of Hermitian operators, it is convenient now to introduce a concept and a symbol which we shall have frequent occasion to use. We shall say that an operator A *commutes* with an operator B, and we shall write $A \leftrightarrow B$, if $AB = BA$.

THEOREM 3. *The product of two Hermitian operators A and B is Hermitian if and only if $A \leftrightarrow B$.*

Proof. Since $(AB)^* = BA$, the equations $(AB)^* = AB$ and $BA = AB$ are obviously equivalent.

From Theorem 3 and the discussion that preceded it we conclude that if A is a Hermitian operator and p is a real polynomial, then $p(A)$ is Hermitian.

The evidence we have collected tends to show (cf. in particular Theorem 1) that if we think of an operator as a generalized complex number, then we should think of a Hermitian operator as a generalized real number. Such an attitude is quite fruitful. It suggests, for instance, that we may define a concept of positiveness for Hermitian operators; we shall, indeed, say that a Hermitian operator A is *positive*, in symbols $A \geqq 0$, if $(Ax, x) \geqq 0$ for every vector x. It is obvious that a positive multiple of a positive operator and the sum of two positive operators are positive. We may continue further along the lines suggested by these considerations and define a partial order in the set of Hermitian operators by writing $A \leqq B$ whenever $B - A$ is positive. This ordering is *proper* (i.e. if $A \leqq B$ and $B \leqq A$, then $A = B$) and *transitive* (i.e. if $A \leqq B$ and $B \leqq C$, then $A \leqq C$). We shall have opportunity to refer to some of these facts later.

§25. Normal and Unitary Operators

If A is any operator, then there exist two uniquely determined Hermitian operators B and C such that $A = B + iC$: in this respect also Hermitian operators imitate the behavior of real numbers. The existence of what might be called the real and the imaginary parts of A is proved by explicitly exhibiting them through the equations $B = \frac{1}{2}(A + A^*)$ and $C = \frac{1}{2i}(A - A^*)$; uniqueness follows from the observation that if $A = B + iC$, then $A^* = B^* - iC^*$.

The fact that in general the real and the imaginary parts of an operator fail to commute is what makes operator theory significantly harder than the corresponding theory of complex numbers and motivates the definition of a normal operator as one for which this pathology does not occur. More explicitly, an operator A is called *normal* if $A \leftrightarrow A^*$; if $A = B + iC$, with B and C Hermitian, then it is easy to see that a necessary and sufficient condition for the normality of A is the relation $B \leftrightarrow C$.

THEOREM 1. *A necessary and sufficient condition that an operator A be normal is that $\| Ax \| = \| A^*x \|$ for every vector x.*

Proof. Since $\| Ax \|^2 = (Ax, Ax) = (A^*Ax, x)$ and, similarly, $\| A^*x \|^2 = (A^*x, A^*x) = (AA^*x, x)$, the identity of the left sides of these relations is equivalent to the identity of their right sides and the latter is equivalent to normality.

One source of the importance of the concept of normality is that many facts about Hermitian operators do not depend on the identity $Ax = A^*x$ but only on the identity $\| Ax \| = \| A^*x \|$, and, in virtue of Theorem 1, all such facts are valid for normal operators.

There is a special class of normal operators of considerable interest, namely the operators U which satisfy the equations $UU^* = U^*U = 1$; such operators are called *unitary*. In the same sense in which Hermitian operators are generalized real numbers, unitary operators are generalized complex numbers of absolute value 1. Observe that a unitary operator is invertible and that in fact unitary operators may be characterized as those invertible operators U for which $U^{-1} = U^*$.

The main reason for the interest of unitary operators is that they are exactly the automorphisms of $\mathfrak{H}$. By an *automorphism* of $\mathfrak{H}$ we mean, of course, an isomorphism from $\mathfrak{H}$ onto $\mathfrak{H}$. Observe that since an isomorphism is an isometry, it follows that an automorphism is in particular an operator.

THEOREM 2. *A necessary and sufficient condition that an operator U be an automorphism of $\mathfrak{H}$ is that it be unitary.*

Proof. Observe that since $(Ux, Uy) = (U^*Ux, y)$, the equation $U^*U = 1$ implies and is implied by the identity $(Ux, Uy) = (x, y)$. Since a unitary operator is invertible and, consequently, is a one-to-one transformation from $\mathfrak{H}$ onto $\mathfrak{H}$, we infer from this observation that a unitary operator is an automorphism. Since (cf. 21.3) an automorphism is also an invertible operator, we infer from the same observation that if U is an automorphism then $U^{-1} = U^*$ and hence that U is unitary

§26. Projections

The *projection* on a subspace $\mathfrak{M}$ is the transformation P defined, for every vector z of the form $x + y$, with $x \in \mathfrak{M}$ and $y \in \mathfrak{M}^{\perp}$, by $Pz = x$.

THEOREM 1. *The projection P on a subspace $\mathfrak{M}$ is an idempotent ($P^2 = P$) and Hermitian ($P^* = P$) operator; if $\mathfrak{M} \neq \mathfrak{O}$, then $\| P \| = 1$.*

Proof. It follows from 13.2 and the projection theorem (whose name is hereby justified) that P is a single-valued transformation from $\mathfrak{H}$ into $\mathfrak{H}$; the fact that P is linear is clear. If $z = x + y$, with $x \in \mathfrak{M}$

and $y \in \mathfrak{M}^{\perp}$, then

$$\| Pz \|^2 = \| x \|^2 \leqq \| x \|^2 + \| y \|^2 = \| z \|^2,$$

so that P is bounded and $\| P \| \leqq 1$. Since $P^2 z = Px = x = Pz$, it follows that P is idempotent. If $\mathfrak{M}$ contains a non-zero vector x, then the fact that $Px = x$ implies that $\| P \| = 1$. If, finally, $z_j = x_j + y_j$, with $x_j \in \mathfrak{M}$ and $y_j \in \mathfrak{M}^{\perp}$, $j = 1, 2$, then

$$(Pz_1, z_2) = (x_1, z_2) = (x_1, x_2) = (z_1, x_2) = (z_1, Pz_2),$$

so that P is Hermitian.

THEOREM 2. *If P is the projection on a subspace $\mathfrak{M}$ and if $\mathfrak{M}_1 = \{x: Px = x\}$ and $\mathfrak{M}_2$ is the range of P, then $\mathfrak{M}_1 = \mathfrak{M}_2 = \mathfrak{M}$.*

Proof. It follows immediately from the definitions of $\mathfrak{M}_1$, $\mathfrak{M}_2$, and P that $\mathfrak{M}_1 \subset \mathfrak{M}_2 \subset \mathfrak{M}$. If, on the other hand, $x \in \mathfrak{M}$, then $Px = x$, so that $\mathfrak{M} \subset \mathfrak{M}_1$ and consequently all these inclusion relations reduce to equalities.

THEOREM 3. *If P is the projection on a subspace $\mathfrak{M}$ and if x is a vector such that $\| Px \| = \| x \|$, then $Px = x$ (and therefore $x \in \mathfrak{M}$).*

Proof. Since $x = Px + (x - Px)$ and since $Px \in \mathfrak{M}$ and $x - Px \in \mathfrak{M}^{\perp}$, it follows that $\| x \|^2 = \| Px \|^2 + \| x - Px \|^2$; the fact that $\| Px \| = \| x \|$ implies therefore that $\| x - Px \| = 0$.

THEOREM 4. *If P is an idempotent Hermitian operator and if $\mathfrak{M}$ is the subspace $\{x: Px = x\}$, then P is the projection on $\mathfrak{M}$.*

Proof. Since P is idempotent, it follows that $P(Pz) = Pz$ for all z; since P is Hermitian, it follows, for every vector x in $\mathfrak{M}$, that $(x, z - Pz) = (x, z) - (Px, z) = 0$ for all z. In other words, $Pz \in \mathfrak{M}$ and $z - Pz \in \mathfrak{M}^{\perp}$ for all z; the theorem follows from the definition of projections and the identity $z = Pz + (z - Pz)$.

We conclude this section with the elementary but exceedingly useful comment that if P is a projection, then $(Px, x) = \| Px \|^2$ for every vector x. The proof of the comment is the following self-explanatory chain of identities:

$$(Px, x) = (PPx, x) = (Px, P^*x) = (Px, Px) = \| Px \|^2.$$

§27. Projections and Subspaces

In view of the results of the preceding section, there is a natural one-to-one correspondence between subspaces and idempotent Hermitian operators. It is in principle possible, therefore, to express all the geometric properties of subspaces in terms of the algebraic properties of their projections. We propose in the following sections to show in detail how that may be done; we begin in the present section by describing the algebraic formulations of invariance, reduction, orthogonal complementation, and orthogonality.

THEOREM 1. *A subspace* $\mathfrak{M}$ *with projection* P *is invariant under an operator* A *if and only if* $AP = PAP$.

Proof. If $AP = PAP$ and if $x \in \mathfrak{M}$, then $Ax = APx = PAPx \in \mathfrak{M}$. If, conversely, $\mathfrak{M}$ is invariant under A, then $APx \in \mathfrak{M}$ and therefore $APx = PAPx$ for every vector x.

THEOREM 2. *A subspace* $\mathfrak{M}$ *with projection* P *reduces an operator* A *if and only if* $P \leftrightarrow A$.

Proof. If $AP = PA$, then, multiplying this relation by P on the right and on the left, we see that both AP and PA are equal to PAP. By the formation of adjoints we obtain the result that both A^*P and PA^* are equal to PA^*P. Since, in view of Theorem 1, the simultaneous validity of the relations $AP = PAP$ and $A^*P = PA^*P$ is equivalent to the assertion that $\mathfrak{M}$ is invariant under both A and A^*, the desired result follows from 23.3.

THEOREM 3. *If* P *is the projection on a subspace* $\mathfrak{M}$, *then* $1 - P$ *is the projection on* $\mathfrak{M}^\perp$ *and* $\mathfrak{M}^\perp = \{x : Px = 0\}$.

Proof. A trivial verification shows that $1 - P$ is idempotent and Hermitian and hence that $1 - P$ is the projection on some subspace $\mathfrak{N}$. By 26.2, $\mathfrak{N} = \{x : (1 - P)x = x\} = \{x : Px = 0\}$; the fact that, therefore, $\mathfrak{N} = \mathfrak{M}^\perp$ follows from the definition of projections.

THEOREM 4. *If* $\mathfrak{M}$ *and* $\mathfrak{N}$ *are subspaces with projections* P *and* Q *respectively, then a necessary and sufficient condition for the validity of all the following relations is the validity of any one of them.*

(i) $\quad \mathfrak{M} \perp \mathfrak{N}$.

(iia) $\quad PQ = 0$.

(iib) $\quad QP = 0$.

(iiia) $\quad P\mathfrak{N} = \mathfrak{O}$.

(iiib) $\quad Q\mathfrak{M} = \mathfrak{O}$.

Proof. If $\mathfrak{M} \perp \mathfrak{N}$, then $\mathfrak{N} \subset \mathfrak{M}^\perp$. Since $Qx \in \mathfrak{N}$ for all x, it follows (by Theorem 3) that $PQx = 0$ for all x. If $PQ = 0$ and if $x \in \mathfrak{N}$, then $Qx = x$ and therefore $Px = PQx = 0$, so that $P\mathfrak{N} = \mathfrak{O}$. If, finally,

$P\mathfrak{N} = \mathfrak{D}$, then (again by Theorem 3) $\mathfrak{N} \subset \mathfrak{M}^{\perp}$ and therefore $\mathfrak{M} \perp \mathfrak{N}$. These arguments prove the equivalence of (i), (iia), and (iiia); the equivalence of these relations to (iib) and (iiib) follows by symmetry. Alternatively we may derive (iia) and (iib) from each other by the consideration of adjoints, and, after observing that (iiia) and (iiib) may be expressed in the form $\mathfrak{N} \subset \mathfrak{M}^{\perp}$ and $\mathfrak{M} \subset \mathfrak{N}^{\perp}$ respectively, derive them from each other by orthogonal complementation.

Justified by Theorem 4 we shall find it convenient to say that two projections P and Q are *orthogonal*, in symbols $P \perp Q$, if $PQ = 0$.

§28. Sums of Projections

In order to discuss the theory of sums of projections in the necessary generality, we have to make a brief digression to describe the concept of not necessarily finite sums of operators. A family $\{A_j\}$ of operators will be called *summable*, and the operator A will be called its *sum*, in symbols $\Sigma_j A_j = A$, if $\Sigma_j A_j x = Ax$ for every vector x. The fact that a scalar factor may be distributed through the terms of a sum, as well as the fact that two sums may be added term by term, follows from the corresponding theorems (7.1 and 7.2) from the theory of summable families of vectors. The fact that, more generally, operator multiplication is distributive with respect to not necessarily finite summation needs a little bit of proof.

THEOREM 1. *If and A and B are operators and if $\{A_j\}$ is a family of operators such that $\Sigma_j A_j = A$, then $\Sigma_j A_j B = AB$ and $\Sigma_j BA_j = BA$.*

Proof. The first assertion is easy: since $\Sigma_j A_j y = Ay$ for every vector y, we may replace y by Bx. The second assertion is easier: from the validity of the relation $\Sigma_j A_j x = Ax$ for every vector x we conclude, from 19.2, that $\Sigma_j BA_j x = BAx$ for all x.

THEOREM 2. *If P is an operator and if $\{P_j\}$ is a family of projections such that $\Sigma_j P_j = P$, then a necessary and sufficient condition that P be a projection is that $P_j \perp P_k$ whenever $j \neq k$, or, in different language, that $\{P_j\}$ be an* ***orthogonal family of projections.*** *If this condition is satisfied and if, for each j, the range of P_j is the subspace $\mathfrak{M}_j$, then the range $\mathfrak{M}$ of P is $\vee_j \mathfrak{M}_j$.*

Proof. If the family $\{P_j\}$ is orthogonal, then

$$P^2 = (\Sigma_j P_j)(\Sigma_k P_k) = \Sigma_j \Sigma_k P_j P_k = \Sigma_j P_j = P$$

and

$$(Px, y) = (\Sigma_j P_j x, y) = \Sigma_j (P_j x, y)$$
$$= \Sigma_j (x, P_j y) = (x, \Sigma_j P_j y) = (x, Py)$$

for every pair of vectors x and y. In other words the orthogonality of the family $\{P_j\}$ implies that P is idempotent and Hermitian, and hence that P is a projection.

If, conversely, P is a projection and if $x \in \mathfrak{M}_k$ for some value of k, then

$$\| x \|^2 \geqq \| Px \|^2 = (Px, x) = \Sigma_j (P_j x, x)$$
$$= \Sigma_j \| P_j x \|^2 \geqq \| P_k x \|^2 = \| x \|^2.$$

It follows that every term in this chain of equations is equal to every other term. From the equality of $\Sigma_j \| P_j x \|^2$ and $\| P_k x \|^2$ we conclude that $P_j x = 0$ whenever $j \neq k$, and hence that $P_j \mathfrak{M}_k = \mathfrak{O}$ whenever $j \neq k$; the orthogonality of the family $\{P_j\}$ follows from 27.4. From the equality of $\| x \|$ and $\| Px \|$ we conclude, by 26.3, that $x \in \mathfrak{M}$ and hence that $\mathfrak{M}_k \subset \mathfrak{M}$ for all k; it follows trivially that $\vee_j \mathfrak{M}_j \subset \mathfrak{M}$. Since, finally, $P_j x \in \mathfrak{M}_j$ for every vector x and every value of j, it follows that $Px = \Sigma_j P_j x \in \Sigma_j \mathfrak{M}_j = \vee_j \mathfrak{M}_j$ for all x, or, since $\mathfrak{M}$ is the range of P, that $\mathfrak{M} \subset \vee_j \mathfrak{M}_j$.

We call attention to the fact that although the proof of Theorem 2 for finite families can be made shorter than the one we presented, its assertion is non-trivial even in that case.

§29. Products and Differences of Projections

The useful fact about products of projections lies near the surface.

THEOREM 1. *A necessary and sufficient condition that the product $P = P_1 P_2$ of two projections P_1 and P_2 be a projection is that $P_1 \leftrightarrow P_2$. If this condition is satisfied and if the ranges of P, P_1, and P_2 are $\mathfrak{M}$, $\mathfrak{M}_1$, and $\mathfrak{M}_2$ respectively, then $\mathfrak{M} = \mathfrak{M}_1 \cap \mathfrak{M}_2$.*

Proof. According to 24.3, P is Hermitian if and only if $P_1 \leftrightarrow P_2$; it is clear that if $P_1 \leftrightarrow P_2$, then P is idempotent. We may already conclude that P is a projection if and only if $P_1 \leftrightarrow P_2$; it remains, assuming that this is the case, to settle the relations among the ranges. Since the range of a product of two operators is obviously contained in the range of the first factor, the commutativity of P_1 and P_2 implies that

$\mathfrak{M} \subset \mathfrak{M}_1$ and $\mathfrak{M} \subset \mathfrak{M}_2$ and therefore that $\mathfrak{M} \subset \mathfrak{M}_1 \cap \mathfrak{M}_2$. If, on the other hand, $x \in \mathfrak{M}_1 \cap \mathfrak{M}_2$, then $P_1x = P_2x = x$ and therefore $Px = x$, so that $\mathfrak{M}_1 \cap \mathfrak{M}_2 \subset \mathfrak{M}$.

Before discussing the facts about differences of projections, we find it convenient to describe the algebraic formulation of the geometric concept of one subspace containing another.

THEOREM 2. *If $\mathfrak{M}$ and $\mathfrak{N}$ are subspaces with projections P and Q respectively, then a necessary and sufficient condition for the validity of all the following relations is the validity of any one of them.*

(i) $$P \leqq Q.$$

(ii) $$\| Px \| \leqq \| Qx \| \textit{ for every vector } x.$$

(iii) $$\mathfrak{M} \subset \mathfrak{N}.$$

(iva) $$QP = P.$$

(ivb) $$PQ = P.$$

Proof. If $P \leqq Q$, then $\| Px \|^2 = (Px, x) \leqq (Qx, x) = \| Qx \|^2$ for every vector x. If $\| Px \| \leqq \| Qx \|$ for all x, and if we consider an arbitrary vector x in $\mathfrak{M}$, then $\| x \| = \| Px \| \leqq \| Qx \| \leqq \| x \|$ (since $\| Q \| \leqq 1$). Since this implies that $\| Qx \| = \| x \|$, it follows from 26.3 that $Qx = x$, i.e. that $x \in \mathfrak{N}$, and hence that $\mathfrak{M} \subset \mathfrak{N}$. If $\mathfrak{M} \subset \mathfrak{N}$, then $Px \in \mathfrak{N}$ and therefore $QPx = Px$ for every vector x. If $QP = P$, then, forming the adjoint of both sides of this relation, we see that $PQ = P$. If $PQ = P$, then

$$(Px, x) = \| Px \|^2 = \| PQx \|^2 \leqq \| Qx \|^2 = (Qx, x)$$

for all x.

THEOREM 3. *A necessary and sufficient condition that the difference $P = P_1 - P_2$ of two projections P_1 and P_2 be a projection is that $P_2 \leqq P_1$. If this condition is satisfied and if the ranges of P, P_1, and P_2 are $\mathfrak{M}$, $\mathfrak{M}_1$, and $\mathfrak{M}_2$ respectively, then $\mathfrak{M} = \mathfrak{M}_1 - \mathfrak{M}_2$.*

Proof. If P is a projection, then

$$(P_1x, x) - (P_2x, x) = (Px, x) = \| Px \|^2 \geqq 0$$

for every vector x. If, conversely, $P_2 \leqq P_1$, then $P_1P_2 = P_2P_1 = P_2$ and therefore

$$(P_1 - P_2)^2 = P_1 - P_1P_2 - P_2P_1 + P_2 = P_1 - P_2.$$

Since P is obviously Hermitian, we may already conclude that P is a projection if and only if $P_2 \leqq P_1$; it remains, assuming that this is the case, to settle the relations among the ranges. Since $P_2 \leqq P_1$ implies that $P_1 \leftrightarrow 1 - P_2$, since $P_1 - P_2 = P_1(1 - P_2)$, and since the range of $1 - P_2$ is $\mathfrak{M}_2^\perp$, it follows from Theorem 1 that

$$\mathfrak{M} = \mathfrak{M}_1 \cap \mathfrak{M}_2^\perp = \mathfrak{M}_1 - \mathfrak{M}_2 .$$

§30. Infima and Suprema of Projections

Not only is there a natural one-to-one correspondence between subspaces and projections, but this correspondence even preserves the relations of order: if $\mathfrak{M}$ and $\mathfrak{N}$ are subspaces with projections P and Q respectively, then a necessary and sufficient condition that $P \leqq Q$ is that $\mathfrak{M} \subset \mathfrak{N}$. It follows (cf. §10) that the set of all projections is a partially ordered set with the property that for any family $\{P_j\}$ of projections there exists a greatest projection (to be denoted by $\bigwedge_j P_j$) smaller than each of them and there exists a smallest projection (to be denoted by $\bigvee_j P_j$) greater than each of them. (For the infimum and supremum of two projections P and Q we shall use the symbols $P \wedge Q$ and $P \vee Q$ respectively.) In other words the partially ordered set of all projections is a complete lattice, an isomorphic copy of the complete lattice of all subspaces. In view of these facts there is a systematic geometric procedure for finding the infimum and the supremum of a family $\{P_j\}$ of projections: if, for each j, the range of P_j is $\mathfrak{M}_j$, then $\bigwedge_j P_j$ is the projection with range $\bigcap_j \mathfrak{M}_j$ and $\bigvee_j P_j$ is the projection with range $\bigvee_j \mathfrak{M}_j$.

It is in general difficult, though not impossible, to describe the infimum and the supremum of a family of projections in algebraic terms. In the presence, however, of suitable orthogonality, or, more generally, commutativity assumptions, the job becomes easy.

THEOREM 1. *If $\{P_j\}$ is an orthogonal family of projections, then* $\bigvee_j P_j = \sum_j P_j$.

Proof. If we knew that the family $\{P_j\}$ were summable, the result would be an immediate consequence of 28.2. Instead of proving summability, however, we find it just as easy to proceed directly. If, for each j, the range of P_j is $\mathfrak{M}_j$ and if the range of $\bigvee_j P_j$ is $\mathfrak{M}$, then $\mathfrak{M} = \bigvee_j \mathfrak{M}_j = \sum_j \mathfrak{M}_j$; cf. 13.2. For an arbitrary vector z write $z = x + y$, with $x \in \mathfrak{M}$ and $y \in \mathfrak{M}^\perp$, and write $x = \sum_j x_j$ with $x_j \in \mathfrak{M}_j$ for all j. Since $P_k x_j = \delta_{jk} x_k$ for all j and k, it follows that $P_k z = \sum_j P_k x_j = x_k$ and hence that $Pz = x = \sum_j x_j = \sum_j P_j z$.

THEOREM 2. *If P_1 and P_2 are two commutative projections, then* $P_1 \wedge P_2 = P_1P_2$ *and* $P_1 \vee P_2 = P_1 + P_2 - P_1P_2$.

Proof. The assertion concerning $P_1 \wedge P_2$ is merely a paraphrase of 29.1. To prove the assertion concerning $P = P_1 + P_2 - P_1P_2$ we introduce the usual notation and denote the ranges of P, P_1, and P_2 by $\mathfrak{M}$, $\mathfrak{M}_1$, and $\mathfrak{M}_2$ respectively. Since $P = P_1 + (1 - P_1)P_2$, it follows that P is a projection and that, in fact, $\mathfrak{M} = \mathfrak{M}_1 \vee (\mathfrak{M}_1^{\perp} \cap \mathfrak{M}_2)$. Since, similarly, $P = P_1(1 - P_2) + P_2$, and since, therefore, $\mathfrak{M} = (\mathfrak{M}_1 \cap \mathfrak{M}_2^{\perp}) \vee \mathfrak{M}_2$, it follows that $\mathfrak{M}_1 \subset \mathfrak{M} \subset \mathfrak{M}_1 \vee \mathfrak{M}_2$ and $\mathfrak{M}_2 \subset \mathfrak{M} \subset \mathfrak{M}_1 \vee \mathfrak{M}_2$. These relations imply that $\mathfrak{M} = \mathfrak{M}_1 \vee \mathfrak{M}_2$ and hence that, indeed, $P = P_1 \vee P_2$.

Our last result along these lines shows that in the presence of commutativity even the sorely missed distributive law is willing to put in an appearance.

THEOREM 3. *If P is a projection and if $\{P_j\}$ is a family of projections such that $P \leftrightarrow P_j$ for all j, then* $P \wedge (\bigvee_j P_j) = \bigvee_j (P \wedge P_j)$.

Proof. Since $P \wedge P_j \leqq P$ and $P \wedge P_j \leqq \bigvee_j P_j$ for all j, it follows that $P \wedge P_j \leqq P \wedge (\bigvee_j P_j)$ for all j and hence that $\bigvee_j (P \wedge P_j) \leqq P \wedge (\bigvee_j P_j)$. This inequality is a lattice-theoretic triviality; to prove that under our assumptions it becomes an equality requires some more work. We shall complete our labors by showing that whenever a vector x belongs to the range of $P \wedge (\bigvee_j P_j)$ and is at the same time orthogonal to the range of $\bigvee_j (P \wedge P_j)$, then that vector x must be 0. In other words we must show that if $x = Px = (\bigvee_j P_j)x$, and if $(\bigvee_j (P \wedge P_j))x = 0$, then $x = 0$. The last-written assumption implies (and here is where we use commutativity) that $P_jPx = 0$ for all j. Since $Px = x$, it follows that $P_jx = 0$, i.e. that x is orthogonal to the range of P_j, for all j. Consequently x is orthogonal to the range of $\bigvee_j P_j$; the only way to reconcile this with the fact that x belongs to that range is to conclude that $x = 0$.

§31. The Spectrum of an Operator

The *spectrum* of an operator A, in symbols $\Lambda(A)$, is the set of all those complex numbers λ for which $A - \lambda$ is not invertible.

The first motivation for considering spectra comes from the finite-dimensional case. If $\mathfrak{H}$ is finite-dimensional, then a necessary and sufficient condition that an operator be not invertible is the vanishing of its determinant—a concept which makes no sense in the general, not necessarily finite-dimensional, case. Since the determinant of $A - \lambda$ is a polynomial in λ, whose zeros are exactly the proper values of A, it follows that in the finite-dimensional case the spectrum of an operator is exactly the set of its proper values.

We recall that the concept of proper value need not be defined in terms of determinants; according to the geometric definition, a complex number λ is a proper value of an operator A if there exists a non-zero vector x such that $Ax = \lambda x$. Equivalently: λ is a proper value of A if there exists a unit vector x such that $\| Ax - \lambda x \| = 0$. This last formulation of the definition admits a reasonable generalization. We shall say that a complex number λ is an *approximate proper value* of an operator A if for every positive number ε there exists a unit vector x such that $\| Ax - \lambda x \| < \varepsilon$; it is easy to verify that an equivalent requirement is that for every positive number ε there exist a non-zero vector x such that $\| Ax - \lambda x \| < \varepsilon \| x \|$. The *approximate point spectrum* of an operator A, in symbols $\Pi(A)$, is the set of approximate proper values of A.

THEOREM 1. *If A is an operator, then $\Pi(A) \subset \Lambda(A)$.*

Proof. If $\lambda \notin \Lambda(A)$, then $A - \lambda$ is invertible and consequently we have

$$\| x \| = \|(A - \lambda)^{-1}(A - \lambda)x \| \leqq \|(A - \lambda)^{-1} \| \cdot \| Ax - \lambda x \|$$

for every vector x. This implies that $\| Ax - \lambda x \| \geqq \varepsilon \| x \|$, with $\varepsilon = 1/\|(A - \lambda)^{-1} \|$, for every vector x, and hence that $\lambda \notin \Pi(A)$.

THEOREM 2. *If A is a normal operator, then $\Pi(A) = \Lambda(A)$.*

Proof. In view of Theorem 1 it is sufficient to prove that $\Lambda(A) \subset \Pi(A)$. If $\lambda \notin \Pi(A)$, then there exists a positive real number ε such that $\| Ay - \lambda y \| \geqq \varepsilon \| y \|$ for every vector y. Since $A - \lambda$ is just as normal as A, and since $(A - \lambda)^* = A^* - \lambda^*$, it follows (cf. 25.1) that $\| A^*y - \lambda^*y \| \geqq \varepsilon \| y \|$ for all y. In order to prove that $\lambda \notin \Lambda(A)$, i.e. that $A - \lambda$ is invertible, it is sufficient, in view of 21.3, to prove that the range of $A - \lambda$ is dense, or, equivalently, that the orthogonal complement of the range is $\mathfrak{O}$. Clearly, however, if a vector y is orthogonal to the range of $A - \lambda$, then $0 = ((A - \lambda)x, y) = (x, (A^* - \lambda^*)y)$ for all x, and hence $A^*y - \lambda^*y = 0$. Since $\| A^*y - \lambda^*y \| \geqq \varepsilon \| y \|$, it follows that $y = 0$ and the proof is complete.

According to Theorems 1 and 2, the spectrum, at least for normal operators, is a more or less natural object. A study of some examples, notably of the multiplication operators described in §20(ii), sheds considerably more light on the subject. By way of illustration we mention, without proof, that the spectrum of the multiplication operator defined by a bounded measurable function h is the essential range of h. By the *essential range* of a complex-valued measurable function h on a measure space with measure μ we mean the set of all those complex numbers λ which have the property that $\mu(h^{-1}(M)) \neq 0$ whenever M is an open set containing λ. This concept is a slight measure-theoretic variant of the usual concept of the range of a function and is not to be confused with the range of the multiplication operator defined by the function; the former is a set of complex numbers and the latter is a set of vectors.

§32. Compactness of Spectra

We begin with an auxiliary result on invertibility.

THEOREM 1. *If an operator A is such that $\|1 - A\| < 1$, then A is invertible.*

Proof. If we write $\|1 - A\| = 1 - \alpha$, so that $0 < \alpha \leqq 1$, then

$$\|Ax\| = \|x - (x - Ax)\| \geqq \|x\| - \|(1 - A)x\| \geqq \|x\| - (1 - \alpha)\|x\| = \alpha\|x\|$$

for every vector x. It follows from 21.3 that it is sufficient, in order to prove the invertibility of A, to show that the range $\mathfrak{M}$ of A is dense in $\mathfrak{H}$. We shall establish the density of $\mathfrak{M}$ by proving that if y is an arbitrary vector, and if $\delta = \inf\{\|y - x\| : x \in \mathfrak{M}\}$, then $\delta = 0$. If $\delta > 0$, then there exists a vector x in $\mathfrak{M}$ such that $(1 - \alpha)\|y - x\| < \delta$. Since $\mathfrak{M}$ contains both x and $A(y - x)$, and therefore also $x + A(y - x)$, it follows that

$$\delta \leqq \|(y - x) - A(y - x)\| \leqq \|1 - A\| \cdot \|y - x\| = (1 - \alpha)\|y - x\| < \delta,$$

and we have reached the desired contradiction.

THEOREM 2. *If A is an operator, then $\Lambda(A)$ is a compact subset of the complex plane; if $\lambda \in \Lambda(A)$, then $|\lambda| \leqq \|A\|$.*

Proof. If $\lambda_0 \,\epsilon'\, \Lambda(A)$, so that $A - \lambda_0$ is invertible, then

$$\|1 - (A - \lambda_0)^{-1}(A - \lambda)\| = \|(A - \lambda_0)^{-1}((A - \lambda_0) - (A - \lambda))\| \leqq \|(A - \lambda_0)^{-1}\| \cdot |\lambda - \lambda_0|$$

and consequently $\| 1 - (A - \lambda_0)^{-1}(A - \lambda)\| < 1$ whenever $|\lambda - \lambda_0|$ is sufficiently small. It follows from Theorem 1 that $(A - \lambda_0)^{-1}(A - \lambda)$ is invertible and therefore that $A - \lambda$ is invertible whenever $|\lambda - \lambda_0|$ is sufficiently small. This implies that the complement of $\Lambda(A)$ is an open subset of the complex plane; it remains only to prove the second assertion of the theorem. If $|\lambda| > \|A\|$, then $\|A/\lambda\| < 1$ and therefore, again by Theorem 1, $1 - (A/\lambda)$ is invertible. It follows that $\lambda \,\epsilon'\, \Lambda(A)$ and hence, contrapositively, that if $\lambda \,\epsilon\, \Lambda(A)$, then $|\lambda| \leqq \|A\|$.

Even in the absence of normality, the approximate point spectrum tries hard to act like the spectrum; as a sample of such behavior we mention that Theorem 2 remains true if Λ is replaced by Π. The proof is easy. If $\lambda_0 \,\epsilon'\, \Pi(A)$, then there exists a positive number ε such that $\|(A - \lambda_0)x\| \geqq \varepsilon$ for all unit vectors x. Consequently if x is a unit vector and if $|\lambda - \lambda_0| < \varepsilon/2$, then

$$\|Ax - \lambda x\| \geqq \|Ax - \lambda_0 x\| - |\lambda_0 - \lambda| \geqq \frac{\varepsilon}{2},$$

so that $\lambda \,\epsilon'\, \Pi(A)$. This means that the complement of $\Pi(A)$ is open; the rest of our assertion is an immediate consequence of 31.1.

§33. Transforms of Spectra

It is interesting to observe what happens to the spectrum of an operator when it is subjected to various elementary transformations. If, for instance, A and B are operators, and if B is invertible, it is easy to see that $\Lambda(A) = \Lambda(B^{-1}AB)$. (In view of the identity $B^{-1}(A - \lambda)B = B^{-1}AB - \lambda$, the invertibility of the right term is equivalent to the invertibility of $A - \lambda$.) In this section we examine the behavior of the spectrum with respect to the formation of polynomials, inverses, and adjoints.

THEOREM 1. *If A is an operator and p is a polynomial, then*

$$\Lambda(p(A)) = p(\Lambda(A)) = \{p(\lambda) : \lambda \,\epsilon\, \Lambda(A)\}.$$

Proof. For any complex number λ_0 there exists a polynomial q such that $p(\lambda) - p(\lambda_0) = (\lambda - \lambda_0)q(\lambda)$ identically in λ. It follows that $p(A) - p(\lambda_0) = (A - \lambda_0)q(A)$; we assert that if $\lambda_0 \,\epsilon\, \Lambda(A)$, then $B = (A - \lambda_0)q(A)$ is not invertible. (If it were, then we should have

$$(A - \lambda_0)\cdot q(A)B^{-1} = BB^{-1} = 1 = B^{-1}B$$
$$= B^{-1}\cdot(A - \lambda_0)q(A) = B^{-1}q(A)\cdot(A - \lambda_0),$$

i.e. $A - \lambda_0$ would also be invertible.) Since this means that $p(A) - p(\lambda_0)$ is not invertible, we have proved that $p(\lambda_0) \in \Lambda(p(A))$ and hence that $p(\Lambda(A)) \subset \Lambda(p(A))$. Suppose on the other hand that $\lambda_0 \in \Lambda(p(A))$, and let $\lambda_1, \cdots, \lambda_n$ be the (not necessarily distinct) roots of the equation $p(\lambda) = \lambda_0$. It follows that $p(A) - \lambda_0 = \alpha(A - \lambda_1) \cdots (A - \lambda_n)$ for a suitable non-zero complex number α, and hence that $A - \lambda_j$ must fail to be invertible for at least one value of j, $1 \leqq j \leqq n$. For such a value of j we have $\lambda_j \in \Lambda(A)$ and $p(\lambda_j) = \lambda_0$, so that $\lambda_0 \in p(\Lambda(A))$ and therefore $\Lambda(p(A)) \subset p(\Lambda(A))$.

THEOREM 2. *If an operator A is invertible, then* $\Lambda(A^{-1}) = (\Lambda(A))^{-1} = \{\lambda^{-1}:\lambda \in \Lambda(A)\}$.

Proof. Observe that since saying that A is invertible is the same as saying that 0 is not in $\Lambda(A)$, the symbol $(\Lambda(A))^{-1}$ makes sense. The identity $A^{-1} - \lambda^{-1} = (\lambda - A)\lambda^{-1}A^{-1}$ shows that if $\lambda \in' \Lambda(A)$, so that $A - \lambda$ is invertible, then $A^{-1} - \lambda^{-1}$ is invertible, so that $\lambda^{-1} \in' \Lambda(A^{-1})$. In other words $\Lambda(A^{-1}) \subset (\Lambda(A))^{-1}$ and our theorem is half proved. The reverse inequality follows by the elegant trick of applying what we have already proved to A^{-1} instead of A.

THEOREM 3. *If A is an operator, then* $\Lambda(A^*) = (\Lambda(A))^* = \{\lambda^*:\lambda \in \Lambda(A)\}$.

Proof. If $\lambda \in' \Lambda(A)$, so that $A - \lambda$ is invertible, then $A^* - \lambda^*$ is invertible, and therefore $\lambda^* \in' \Lambda(A^*)$. Since this proves that $\Lambda(A^*) \subset (\Lambda(A))^*$, the proof may be completed just as in Theorem 2; to obtain the reverse inequality it is sufficient to apply the inequality already proved to A^* instead of A.

§34. The Spectrum of a Hermitian Operator

If 33.3 is applied to a Hermitian operator, it yields the result that the spectrum of a Hermitian operator is symmetric with respect to the real axis. Actually the situation is much simpler.

THEOREM 1. *If A is a Hermitian operator, then $\Lambda(A)$ is a subset of the real axis.*

Proof. If λ is not real, then, for every non-zero vector x,

$$0 < |\lambda - \lambda^*|\cdot\|x\|^2 = |((A - \lambda)x, x) - ((A - \lambda^*)x, x)|$$
$$= |((A - \lambda)x, x) - (x, (A - \lambda)x)| \leqq 2\|Ax - \lambda x\|\cdot\|x\|;$$

the desired result follows from the fact (cf. 31.2) that for Hermitian operators the approximate point spectrum and the spectrum are the same.

Our next result is one of the most powerful tools for the study of Hermitian operators; it asserts that the norm of such an operator can be calculated from its spectrum.

THEOREM 2. *If A is a Hermitian operator, then* $\| A \| = \alpha = \sup \{ |\lambda| : \lambda \in \Lambda(A) \}$.

Proof. The fact that $\alpha \leqq \| A \|$, does not depend on the Hermitian character of A; it follows from 32.2. We shall prove that equality prevails by showing that $\| A \|^2 \in \Pi(A^2)$; in view of 31.1 and 33.1 we shall then be able to conclude that $\pm \| A \| \in \Lambda(A)$ for a suitable choice of the ambiguous sign. The proof of the promised relation is based on the identity

$$\| A^2 x - \lambda^2 x \|^2 = \| A^2 x \|^2 - 2\lambda^2 \| Ax \|^2 + \lambda^4 \| x \|^2,$$

valid (since A is Hermitian) for all real numbers λ and all vectors x. If $\{x_n\}$ is a sequence of unit vectors such that $\| Ax_n \| \rightarrow \| A \|$, and if $\lambda = \| A \|$, then it follows from our identity that

$$\| A^2 x_n - \lambda^2 x_n \|^2 \leqq (\| A \| \cdot \| Ax_n \|)^2 - 2\lambda^2 \| Ax_n \|^2 + \lambda^4$$

$$= \lambda^4 - \lambda^2 \| Ax_n \|^2 \rightarrow 0$$

and hence that we do indeed have $\| A \|^2 \in \Pi(A^2)$.

One of the useful conclusions we can draw from Theorem 2 is that the spectrum of a Hermitian operator is not empty. This is not a trivial conclusion. We shall obtain the corresponding fact for normal operators only after the application of a lot more relatively deep analysis. We hereby report that the spectrum of an arbitrary operator is also not empty; since we shall have no occasion to make use of this fact, we shall not enter into its proof.

To state our last result, an easy corollary of Theorem 2, we introduce some new notation. If A is an operator and if f is a complex-valued function on the spectrum of A, we shall write

$$\mathbf{N}_A(f) = \sup \{ |f(\lambda)| : \lambda \in \Lambda(A) \}.$$

THEOREM 3. *If A is a Hermitian operator and p is a real polynomial, then* $\| p(A) \| = \mathbf{N}_A(p)$.

Proof. Applying first Theorem 2 (to $p(A)$ instead of A) and then 33.1, we obtain

$$\begin{aligned} \| p(A)\| &= \sup \{|\lambda|:\lambda \in \Lambda(p(A))\} \\ &= \sup \{|\lambda|:\lambda \in p(\Lambda(A))\} \\ &= \sup \{|p(\lambda)|:\lambda \in \Lambda(A)\}. \end{aligned}$$

§35. Spectral Heuristics

We are now in a position to make a deep analysis of the structure of Hermitian and, more generally, normal operators. In order, however, to motivate and illustrate not only the method of proof but even the statement of the facts, it is advisable that we make a brief digression and examine an analogous but more elementary theory.

Consider the statement that a real-valued, bounded, measurable function f on a finite measure space X can be uniformly approximated by simple functions. More precisely: to any positive number ε there corresponds a finite, disjoint family of measurable sets, or, equivalently, a finite, disjoint family $\{\chi_j\}$ of measurable characteristic functions, and a finite family $\{\lambda_j\}$ of real numbers, such that $|f(t) - \Sigma_j\lambda_j\chi_j(t)| < \varepsilon$ for all t in X.

How does the usual proof of this theorem go? If the bounds of f are α and β, so that $\alpha \leqq f(t) \leqq \beta$ for all t in X, we may subdivide the interval $[\alpha, \beta]$ into a finite, disjoint family $\{M_j\}$ of intervals of length less than ε, and, for each j, we may select a number λ_j in M_j. In the subset $f^{-1}(M_j)$ of X the values of f are all within ε of λ_j, and therefore we obtain the desired result by setting χ_j equal to the characteristic function of $f^{-1}(M_j)$. (Note that since the value at a point t of the characteristic function of $f^{-1}(M_j)$ is equal to the value of the characteristic function of M_j at $f(t)$, we have $\chi_j(t) = \chi_{M_j}(f(t))$ for all t.) If, for any Borel set M in the real line, we write $E(M)$ for the characteristic function of the subset $f^{-1}(M)$ of X, our result may be expressed by writing

$$|f - \Sigma_j\lambda_jE(M_j)| < \varepsilon.$$

The expression $\Sigma_j\lambda_jE(M_j)$ looks suspiciously like the sort of sum that occurs in various approaches to integration. The function E is a set function, a measure in some sense, which associates a certain characteristic function on the space X with each Borel set in the real line. Since, for each j, λ_j is a point in the element M_j of a certain partition of the interval $[\alpha, \beta]$, the integral that appears to be lurking in the background has the form $\int \lambda\, dE(\lambda)$. It is not a difficult task to construct a theory of integration in which symbols such as $\int \lambda\, dE(\lambda)$ make sense, although of course our heuristic hints do not constitute such a construction.

Proceeding formally, we may summarize our comments as follows. The approximability of a real-valued, bounded, measurable function f by simple functions can be expressed by writing $f = \int \lambda\, dE(\lambda)$, where E is the somewhat peculiar, function-valued, "measure" whose value at a Borel set M in the real line is the characteristic function of $f^{-1}(M)$. The measure E has some unusual properties and reflects in some interesting ways the structure of the function f. Among its properties we mention its idempotence ($(E(M))^2 = E(M)$ for every Borel set M) and, more generally, its multiplicativity ($E(M \cap N) = E(M)E(N)$ for every pair of Borel sets M and N). The way in which E reflects the properties of f is illustrated by the assertion that, if M is a Borel set, a necessary and sufficient condition for the vanishing of $E(M)$ is that M be disjoint from the range of f.

The analogs of bounded, real-valued, measurable functions in Hilbert space theory are bounded, Hermitian, linear transformations, i.e. Hermitian operators. Since a function is the characteristic function of a set if and only if it is idempotent, it is clear on algebraic grounds that the analogs of characteristic functions are projections. The approximability of functions by simple functions corresponds in the analogy to the approximability of Hermitian operators by real, finite linear combinations of projections. The purpose of such an operatorial approximation theorem is, just as in the analogous functional situation, to provide a tool for deriving and understanding the deep structural properties of complicated objects in terms of simple objects. For a Hermitian operator, just as for a real function, we shall be able to construct a "measure" E with the multiplicative property mentioned in the preceding paragraph and to recapture the operator by means of an integral. The measure E will reflect the properties of the given operator in many ways; in analogy with our remarks concerning the range of a function, for instance, it will be easy to characterize the spectrum of the operator in terms of E.

The theory for complex-valued, bounded, measurable functions is no harder than for real functions. The proper analog of a complex function turns out to be not any old operator but a normal operator; it will be technically convenient to derive the complex (normal) generalization from the real (Hermitian) special case.

It is customary to motivate the theory we intend to develop not by such analytic considerations as we have indicated, but by reference to the algebraic facts concerning operators on finite-dimensional spaces. It is a good idea to keep both in mind, and, specifically, the reader is advised to think through the relation between our past and future comments on the one hand and the familiar reduction of a Hermitian matrix to diagonal form on the other hand.

§36. Spectral Measures

If X is a set with a specified Boolean σ-algebra **S** of subsets, a *spectral measure* in X is a function E whose domain is **S** and whose values are idempotent, Hermitian operators (projections) on $\mathfrak{H}$, such that $E(X) = 1$ and such that $E(\bigcup_n M_n) = \Sigma_n E(M_n)$ whenever $\{M_n\}$ is a disjoint sequence of sets in **S**. A set X with a specified Boolean σ-algebra **S** of subsets is usually called a *measurable space* and is denoted by $(X, \mathbf{S})$; the sets belonging to **S** are called the *measurable subsets* of X. A typical example of a spectral measure is obtained by letting X be not only a measurable space but a measure space with measure μ, considering the Hilbert space $\mathfrak{L}_2(\mu)$ in the role of $\mathfrak{H}$, and writing $E(M)f = \chi_M f$ whenever $M \in \mathbf{S}$ and $f \in \mathfrak{L}_2(\mu)$ (where χ_M denotes, of course, the characteristic function of the set M). The standard techniques of elementary measure theory show that if E is a spectral measure, then $E(0) = 0$ and E is finitely additive (i.e. $E(\bigcup_j M_j) = \Sigma_j E(M_j)$ whenever $\{M_j\}$ is a finite disjoint family of measurable sets).

THEOREM 1. *If E is a finitely additive, projection-valued set function on the class* **S** *of all measurable subsets of a measurable space (in particular if E is a spectral measure), then E is* **monotone** *and* **subtractive**, *i.e. if M and N are in* **S** *and $M \subset N$, then $E(M) \leqq E(N)$ and $E(N - M) = E(N) - E(M)$.*

Proof. Since $E(N) = E(M) + E(N - M)$, the fact that E is subtractive is trivial; monotony follows from 29.3.

THEOREM 2. *If E is a finitely additive, projection-valued set function on the class* **S** *of all measurable subsets of a measurable space (in particular if E is a spectral measure), then E is* **modular** *and* **multiplicative**, *i.e. if M and N are in* **S**, *then*

$$E(M \cup N) + E(M \cap N) = E(M) + E(N)$$

and

$$E(M \cap N) = E(M)E(N).$$

Proof. If we add $E(M \cap N)$ to both sides of the equation

$$E(M \cup N) = E(M - N) + E(M \cap N) + E(N - M),$$

we obtain

$$E(M \cup N) + E(M \cap N) = (E(M - N) + E(M \cap N)) + (E(N - M) + E(M \cap N)) = E(M) + E(N).$$

This already proves modularity. Since, by Theorem 1, $E(M \cap N) \leqq E(M) \leqq E(M \cup N)$, it follows that $E(M)E(M \cap N) = E(M \cap N)$ and $E(M)E(M \cup N) = E(M)$. If, therefore, we multiply both sides of the modular equation by $E(M)$, we obtain $E(M) + E(M \cap N) = E(M) + E(M)E(N)$, and this proves that E is multiplicative.

We remark that the multiplicative property of E implies in particular that $E(M) \leftrightarrow E(N)$ whenever M and N are in S.

THEOREM 3. *A projection-valued function E on the class S of measurable subsets of a measurable space X is a spectral measure if and only if*

(i) $$E(X) = 1,$$

and

(ii) for each pair of vectors x and y, the complex-valued set function μ defined for every M in S by $\mu(M) = (E(M)x, y)$ is countably additive.

Proof. If E is a spectral measure, then (i) holds by definition and (ii) follows from the fact (7.3) that an inner product one factor of which is an infinite sum may be formed term by term. Suppose, conversely, that (i) and (ii) hold. If M and N are disjoint measurable sets, then the identity

$$(E(M \cup N)x, y) = (E(M)x, y) + (E(N)x, y) = ((E(M) + E(N))x, y)$$

proves that $E(M \cup N) = E(M) + E(N)$, i.e. that E is finitely additive (and therefore multiplicative). If, similarly, $\{M_n\}$ is a disjoint sequence of measurable sets with $\bigcup_n M_n = M$, it is tempting to argue that

$$(E(M)x, y) = \Sigma_n(E(M_n)x, y) = ((\Sigma_n E(M_n))x, y)$$

for all x and y, and hence that $E(M) = \Sigma_n E(M_n)$. The only thing wrong with this argument is that $\Sigma_n E(M_n)$ need not make sense; we shall finish the proof by showing that it does. The multiplicativity of E implies that $\{E(M_n)\}$ is an orthogonal sequence of projections and hence that $\{E(M_n)x\}$ is an orthogonal sequence of vectors for every x.

Since

$$\Sigma_n \| E(M_n)x \|^2 = \Sigma_n(E(M_n)x, x) = (E(M)x, x) = \| E(M)x \|^2,$$

it follows that the sequence $\{E(M_n)x\}$ is summable. If $\Sigma_n E(M_n)x = Ax$, then it is clear that A is a linear transformation of $\mathfrak{H}$ into itself; the chain of equations used to prove the existence of A implies also that A is bounded (and, in fact, that $\| A \| \leqq 1$).

§37. Spectral Integrals

Throughout this section we shall work with an arbitrary but fixed measurable space $(X, \mathbf{S})$; the expression "spectral measure" will always refer to a spectral measure in X. It will be convenient to use also the symbol $\mathfrak{B}$ for the class of all complex-valued, bounded, measurable functions on X, and to write $\mathbf{N}(f) = \sup \{|f(\lambda)|:\lambda \in X\}$ whenever $f \in \mathfrak{B}$.

THEOREM 1. *If E is a spectral measure and if $f \in \mathfrak{B}$, then there exists a unique operator A such that $(Ax, y) = \int f(\lambda)\, d(E(\lambda)x, y)$ for every pair of vectors x and y; the dependence of A on f and E will be denoted by writing $A = \int f\, dE = \int f(\lambda)\, dE(\lambda)$.*

Proof. The boundedness of f implies that the integral $\varphi(x, y) = \int f(\lambda)\, d(E(\lambda)x, y)$ may be formed for every pair of vectors x and y; an obvious computation shows that φ is a bilinear functional. Since $|\varphi(x, x)| \leqq \int |f(\lambda)|\, d \| E(\lambda)x \|^2 \leqq \mathbf{N}(f) \cdot \| x \|^2$, it follows, by 18.2, that φ is bounded and hence, by 22.1, that there does indeed exist a unique operator satisfying the conditions required of A.

THEOREM 2. *If E is a spectral measure, if f and g are in $\mathfrak{B}$, and if α is a complex number, then*

$$\int (\alpha f)\, dE = \alpha \int f\, dE, \qquad \int (f + g)\, dE = \int f\, dE + \int g\, dE,$$

and

$$\int f^*\, dE = (\int f\, dE)^*.$$

Proof. The proofs of all three assertions are similar and almost automatic. To prove, for instance, the last one, we write $A = \int f\, dE$ and $B = \int f^*\, dE$, and we observe that the relations

$$\begin{aligned}(x, By) &= (By, x)^* = (\int f^*(\lambda)\, d(E(\lambda)y, x))^* \\ &= \int f(\lambda)\, d(x, E(\lambda)y) = \int f(\lambda)\, d(E(\lambda)x, y) = (Ax, y)\end{aligned}$$

are valid for every pair of vectors x and y.

THEOREM 3. *If E is a spectral measure and if f and g are in $\mathfrak{B}$, then* $(\int f\,dE)(\int g\,dE) = \int fg\,dE$.

Proof. We write $A = \int f\,dE$ and $B = \int g\,dE$. If the (complex) measure μ in X is defined for every set M in **S** by $\mu(M) = (E(M)Bx, y)$, where x and y are any fixed vectors, then

$$\mu(M) = (Bx, E(M)y) = \int g(\lambda)\,d(E(\lambda)x, E(M)y) = \int g(\lambda)\,d(E(M)E(\lambda)x, y)$$
$$= \int g(\lambda)\,d(E(M \cap \lambda)x, y) = \int_M g(\lambda)\,d(E(\lambda)x, y)$$

for every M in **S**. It follows that

$$(ABx, y) = (A^*y, Bx)^* = (\int f^*(\lambda)\,d(E(\lambda)y, Bx))^*$$
$$= (\int f^*(\lambda)\,d(y, E(\lambda)Bx))^* = \int f(\lambda)\,d(E(\lambda)Bx, y)$$
$$= \int f(\lambda)\,d\mu(\lambda) = \int f(\lambda)g(\lambda)\,d(E(\lambda)x, y)$$

and hence that $AB = \int fg\,dE$.

It follows from the preceding results that if E is a spectral measure, then $\int dE(\lambda) = E(X) = 1$, and, more generally, $\int \chi_M(\lambda)\,dE(\lambda) = \int_M dE(\lambda) = E(M)$ for every M in **S** (Theorem 1); if f and g are in $\mathfrak{B}$, then $\int f\,dE \leftrightarrow \int g\,dE$ (Theorem 3); and if $f \in \mathfrak{B}$, then $\int f\,dE$ is normal (Theorem 2 and the commutativity result just mentioned). To state our last result concerning the algebraic behavior of spectral integrals, we introduce a convenient notation: if E is a spectral measure and B is an operator, we shall write $E \leftrightarrow B$ for the assertion that $E(M) \leftrightarrow B$ for all M in **S**. We remark, for example, that if $f \in \mathfrak{B}$, then $E \leftrightarrow \int f\,dE$.

THEOREM 4. *If E is a spectral measure, if B is an operator such that $E \leftrightarrow B$, and if $f \in \mathfrak{B}$, then $\int f\,dE \leftrightarrow B$.*

Proof. If $\int f\,dE = A$, then

$$(ABx, y) = \int f(\lambda)\,d(E(\lambda)Bx, y) = \int f(\lambda)\,d(BE(\lambda)x, y)$$
$$= \int f(\lambda)\,d(E(\lambda)x, B^*y) = (Ax, B^*y) = (BAx, y)$$

for every pair of vectors x and y.

§38. Regular Spectral Measures

Throughout this section we shall assume that X is a locally compact Hausdorff space and that **S** is the σ-algebra of all Borel sets in X; except for this specialization, we continue to follow the conventions of the preceding section.

A spectral measure E is *regular* if $E(M_0) = \vee E(M)$ for every Borel set M_0, where the supremum is extended over all compact sets M contained in M_0. The *spectrum* of a spectral measure E, in symbols $\Lambda(E)$, is the complement in X of the union of all those open sets M for which $E(M) = 0$. A spectral measure is *compact* if its spectrum is compact. We observe, concerning these definitions, that they cannot even be formulated, let alone exemplified, if X is not a topological space. On the other hand as soon as X is a topological space these definitions make sense; we restrict attention to the case of locally compact Hausdorff spaces mainly because that is the limit of the generality we need for any of our applications.

THEOREM 1. *If E is a regular spectral measure and $\Lambda = \Lambda(E)$, then Λ is a closed set such that $E(X - \Lambda) = 0$ (and therefore $E(\Lambda) = 1$).*

Proof. Since $X - \Lambda$ is, by definition, a union of open sets, Λ is closed. To prove that $E(X - \Lambda) = 0$, it is, in view of regularity, sufficient to prove that $E(M) = 0$ whenever M is a compact subset of $X - \Lambda$. The definition of the spectrum of E implies that every point of $X - \Lambda$, and therefore in particular every point of M, is contained in an open set on which the value of E vanishes. Since M is compact, M may be covered by a finite number of such open sets, and it follows indeed that $E(M) = 0$.

It is frequently convenient to consider spectral integrals such as $\int f(\lambda)\, dE(\lambda)$ even if the complex-valued measurable function f is not bounded; the theory of such integrals remains simple as long as we assume that f is, so to speak, bounded with respect to the regular spectral measure E. More precisely what is needed is that f be bounded on the spectrum Λ of E. If that is the case we define $\int f\, dE$ to mean $\int_\Lambda f\, dE = \int \chi_\Lambda f\, dE$; in view of Theorem 1 this definition will lead to a consistent theory. Another way of accomplishing the same purpose is to replace the space X by the subset Λ and the spectral measure E in X by the spectral measure in Λ obtained by restricting the domain of definition of E to Borel subsets of Λ only. In connection with this circle of ideas it is natural to write $\mathbf{N}_E(f) = \sup\{|f(\lambda)| : \lambda \in \Lambda(E)\}$ whenever E is a spectral measure and f is a complex-valued measurable function bounded on $\Lambda(E)$.

THEOREM 2. *If E is a compact and regular spectral measure with spectrum Λ and if f is a complex-valued continuous function on X, then $\| \int f\, dE \| = \mathbf{N}_E(f)$.*

Proof. We write $\int f\, dE = A$, and we assume first that f is real.

Since it follows from the preceding section that A is Hermitian, we have

$$\| A \| = \sup \{|(Ax, x)| : \| x \| = 1\}.$$

Since, however,

$$|(Ax, x)| \leqq \int_\Lambda |f(\lambda)| \, d \| E(\lambda)x \|^2 \leqq \mathbf{N}_E(f) \cdot \| x \|^2$$

for every vector x, it follows that $\| A \| \leqq \mathbf{N}_E(f)$.

If $\mathbf{N}_E(f) \neq 0$, let ε be a positive number such that $\varepsilon < \mathbf{N}_E(f)$. We may and do assume, without any loss of generality, that $\mathbf{N}_E(f) = \sup \{f(\lambda) : \lambda \in \Lambda\}$. If $M = \{\lambda : f(\lambda) > \mathbf{N}_E(f) - \varepsilon\}$, then M is an open set and $M \cap \Lambda \neq 0$; it follows that $E(M) \neq 0$. If x is a non-zero vector in the range of $E(M)$, then $E(X - M)x = E(X)x - E(M)x = 0$ and therefore

$$|(Ax, x)| = |\int f(\lambda) \, d(E(\lambda)x, x)| = | \int_M f(\lambda) \, d \| E(\lambda)x \|^2 |$$
$$\geqq (\mathbf{N}_E(f) - \varepsilon) \cdot \| x \|^2.$$

It follows that $\| A \| \geqq \mathbf{N}_E(f) - \varepsilon$ for every positive number ε and hence that $\| A \| \geqq \mathbf{N}_E(f)$.

If f is complex, then, by 22.4,

$$\| A \|^2 = \| A^*A \| = \|(\int f^* \, dE)(\int f \, dE)\| = \| \int f^*f \, dE \| .$$

Since $f^*f = |f|^2$ is real, we have

$$\| A \|^2 = \sup \{|f(\lambda)|^2 : \lambda \in \Lambda\} = \mathbf{N}_E(|f|^2) = (\mathbf{N}_E(f))^2.$$

§39. Real and Complex Spectral Measures

A spectral measure defined on the class of all Borel sets of the complex plane is called a *complex spectral measure*. Our first result is that the results of the preceding section are applicable to complex spectral measures.

THEOREM 1. *Every complex spectral measure is regular.*

Proof. The proof of this theorem may be carried out by imitating the proof of the corresponding fact for ordinary numerical measures. The main tool of that proof is the separability of the complex plane. As a compromise between reproducing here all the details of a standard technique on the one hand and saying that the proof is left as an exercise for the reader on the other hand, we shall reduce the theorem as stated to the numerical case. Suppose then that E is a complex spectral measure and that M_0 is a Borel set in the complex plane. If M is a compact subset of M_0, then $E(M_0) \geqq E(M)$ and consequently

$$E(M_0) \geqq \vee E(M).$$

We must show that if a vector x in the range of $E(M_0)$ is orthogonal to the range of $E(M)$ for every compact subset M of M_0, then $x = 0$. If, however, $\mu(M) = (E(M)x, x)$ for every Borel set M, then, by the regularity of numerical measures, $\mu(M_0) = \sup \mu(M)$, where, again, the supremum is extended over all compact subsets of M_0. Since, by hypothesis, $\mu(M) = 0$ for each such compact set, it follows that

$$\mu(M_0) = 0$$

and hence that $x = E(M_0)x = 0$.

Our next result is the reason and justification for using the word "spectrum" in connection with spectral measures.

THEOREM 2. *If E is a compact, complex spectral measure and if $A = \int \lambda \, dE(\lambda)$, then $\Lambda(E) = \Lambda(A)$.*

Proof. If $\lambda_0 \epsilon' \Lambda(E)$, then there is an open set M such that $\lambda_0 \epsilon M$ and $E(M) = 0$. If M' is the complement of M and δ is the distance between λ_0 and M', then

$$\| Ax - \lambda_0 x \|^2 = ((A - \lambda_0)^*(A - \lambda_0)x, x)$$
$$= \int (\lambda - \lambda_0)^*(\lambda - \lambda_0) \, d(E(\lambda)x, x)$$

for every vector x. Since $E(M) = 0$, it follows that

$$\| Ax - \lambda_0 x \|^2 = \int_{M'} | \lambda - \lambda_0 |^2 d(E(\lambda)x, x) \geqq \delta^2 \| x \|^2$$

for all x and hence that

$$\lambda_0 \epsilon' \Pi(A) = \Lambda(A).$$

If, conversely, $\lambda_0 \epsilon \Lambda(E)$, then we have $E(M) \neq 0$ for every open set M containing λ_0. Hence if δ is any positive number and

$$M = \{\lambda : | \lambda - \lambda_0 | < \delta\},$$

then the range of $E(M)$ contains some unit vector x. Since, arguing as before, $\| Ax - \lambda_0 x \|^2 = \int_M | \lambda - \lambda_0 |^2 d(E(\lambda)x, x) \leqq \delta^2$, it follows that $\lambda_0 \epsilon \Lambda(A)$.

A spectral measure defined on the class of all Borel sets of the real line is called a *real spectral measure*. It follows from Theorem 1 (cf. 38.1) that a complex spectral measure whose spectrum is contained in the real axis may be viewed as a real spectral measure; conversely, of course, every real spectral measure may be viewed as a complex spectral measure whose spectrum is contained in the real axis. Consequently any result valid for all complex spectral measures is valid for all real spectral measures as well.

§40. Complex Spectral Integrals

THEOREM 1. *If* E_1 *and* E_2 *are compact, complex spectral measures such that* $\int \lambda dE_1(\lambda) = \int \lambda dE_2(\lambda)$, *then* $E_1 = E_2$.

Proof. Let $\beta(\lambda)$ and $\gamma(\lambda)$ be the real and imaginary part respectively of the complex number λ. If for an arbitrary but fixed vector x we write $\mu_1(M) = (E_1(M)x, x)$ and $\mu_2(M) = (E_2(M)x, x)$, then, since μ_1 and μ_2 are real (and in fact non-negative), it follows that $\int \beta d\mu_1 = \int \beta d\mu_2$ and $\int \gamma d\mu_1 = \int \gamma d\mu_2$. By polarization we obtain the result that

$$\int \beta dE_1 = \int \beta dE_2$$

and

$$\int \gamma dE_1 = \int \gamma dE_2 .$$

The additive and multiplicative properties of spectral integrals imply that if p is any real polynomial in two variables, then

$$\int p(\beta(\lambda), \gamma(\lambda)) d(E_1(\lambda)x, y) = \int p(\beta(\lambda),\gamma(\lambda)) d(E_2(\lambda)x, y)$$

for every pair of vectors x and y. It follows that

$$(E_1(M)x, y) = (E_2(M)x, y)$$

for every Borel set M and all x and y, and consequently $E_1 = E_2$.

Theorem 1 says that a compact, complex spectral measure is uniquely determined by one of the simplest spectral integrals that can be formed, i.e. the integral of the function f defined for every complex number λ by $f(\lambda) = \lambda$. Since it is true (cf. our heuristic promise in §35 and its fulfillment in §44) that every normal operator has the form $\int \lambda dE(\lambda)$ for a suitable compact, complex spectral measure E, Theorem 1 is the assertion that the representation of a normal operator by such an integral is unique.

THEOREM 2. *If* E *is a compact, complex spectral measure and if* B *is an operator such that both* $\int \lambda dE(\lambda)$ *and* $\int \lambda^* dE(\lambda)$ *commute with* B, *then* $E \leftrightarrow B$.

Proof. We use the notation established in the proof of Theorem 1. Since $\beta(\lambda) = \frac{1}{2}(\lambda + \lambda^*)$ and $\gamma(\lambda) = \frac{1}{2i}(\lambda - \lambda^*)$, our assumptions imply that if p is any real polynomial in two variables and if

$$A = \int p(\beta(\lambda), \gamma(\lambda))\, dE(\lambda),$$

then $B \leftrightarrow A$. It follows that

$$\int p(\beta(\lambda), \gamma(\lambda))\, d(E(\lambda)x, B^*y) = (Ax, B^*y) = (ABx, y)$$
$$= \int p(\beta(\lambda), \gamma(\lambda))\, d(E(\lambda)Bx, y)$$

for every pair of vectors x and y. Since we may infer that

$$(BE(M)x, y) = (E(M)x, B^*y) = (E(M)Bx, y)$$

holds identically in the Borel set M and the vectors x and y, the proof is complete.

Theorems 1 and 2 are of course true for real spectral measures in particular; the proofs for this special case are slightly easier than the ones we presented. We observe also that even the statement of Theorem 2 becomes simpler if the spectral measure E is real, since in that case the vanishing of E on the complement of the real axis implies that

$$\int \lambda\, dE(\lambda) = \int \lambda^*\, dE(\lambda).$$

In other words if E is a compact, real spectral measure and if B is an operator such that $\int \lambda\, dE(\lambda) \leftrightarrow B$, then $E \leftrightarrow B$. It is a remarkable and useful fact that this strengthened version of Theorem 2 is true for complex spectral measures also, but it will take us all the work of the following two sections to prove that.

We end this section by reminding the reader of the existence of 37.4. That theorem shows that whenever we have accumulated enough assumptions to justify the conclusion of Theorem 2, then we may also conclude that $\int f\, dE \leftrightarrow B$ for every complex-valued, measurable function f which is bounded on the entire complex plane, or at any rate on the spectrum of the spectral measure E.

§41. Description of the Spectral Subspaces

THEOREM 1. *If A is a normal operator and if $\mathfrak{F} = \mathfrak{F}(A)$ is the set of all vectors x such that $\| A^n x \| \leqq \| x \|$ for every positive integer n, then $\mathfrak{F}$ is a subspace. If B is an operator such that $A \leftrightarrow B$, then $\mathfrak{F}$ is invariant under B.*

Proof. Let $\mathfrak{M}$ be the set of all those vectors x for which the sequence $\{\| A^n x \|\}$ is bounded. If x and y are in $\mathfrak{M}$ and if α and β are complex numbers, then the relation

$$\| A^n(\alpha x + \beta y) \| \leqq | \alpha | \cdot \| A^n x \| + | \beta | \cdot \| A^n y \| ,$$

valid for every positive integer n, shows that $\alpha x + \beta y \in \mathfrak{M}$. If $x \in \mathfrak{M}$, then the relation $\| A^n Bx \| = \| BA^n x \| \leqq \| B \| \cdot \| A^n x \|$, valid for every positive integer n, shows that $Bx \in \mathfrak{M}$. In other words $\mathfrak{M}$ is a linear manifold and $\mathfrak{M}$ is invariant under B; it is clear that $\mathfrak{F} \subset \mathfrak{M}$. It is not at all obvious that $\mathfrak{M}$ is a subspace (i.e. that $\mathfrak{M}$ is closed) and, although the fact that $\mathfrak{F}$ is closed is easy to see, it is not at all obvious that $\mathfrak{F}$ is a linear manifold nor that $\mathfrak{F}$ has the desired invariance property. All these difficulties can be cleared up in one fell swoop by showing that $\mathfrak{F} = \mathfrak{M}$; that is what we propose to do. It is sufficient to prove that if a vector x is such that $\| A^m x \| > \alpha \| x \|$ for some positive integer m and for some number α, $\alpha > 1$, then the sequence $\{\| A^n x \|\}$ is not bounded. But this is easy: an inductive repetition of the argument used to prove the chain of relations

$$\begin{aligned} \alpha^2 \| x \|^2 < \| A^m x \|^2 &= (A^m x, A^m x) = ((A^m)^* A^m x, x) \\ &\leqq \| (A^m)^* A^m x \| \cdot \| x \| = \| A^{2m} x \| \cdot \| x \| \end{aligned}$$

shows that $\| A^{2^n m} x \| > \alpha^{2^n} \| x \|$ for every positive integer n. (The normality of A was used, via 25.1, in the step $\| (A^m)^* A^m x \| = \| A^{2m} x \|$.)

Suppose now that E is a compact, complex spectral measure and that A is the normal operator $\int \lambda dE(\lambda)$. For each complex number λ and each positive real number ε we shall write $\mathfrak{F}(\lambda, \varepsilon)$ for the subspace $\mathfrak{F}\left(\frac{1}{\varepsilon}(A - \lambda)\right)$ associated with the normal operator $\frac{1}{\varepsilon}(A - \lambda)$ in the manner described by Theorem 1. More explicitly $\mathfrak{F}(\lambda, \varepsilon)$ is the set (subspace) of all vectors x such that $\| (A - \lambda)^n x \| \leqq \varepsilon^n \| x \|$ for every positive integer n; roughly speaking a vector x in $\mathfrak{F}(\lambda, \varepsilon)$ may be described as an approximate proper vector with proper value λ and degree of approximation ε. (Use of this language is not to be confused, however, with the technical term defined in §31.) For every set M of complex numbers and for every positive real number ε, we shall write $\mathfrak{F}(M, \varepsilon) = \bigvee_{\lambda \in M} \mathfrak{F}(\lambda, \varepsilon)$, and $\mathfrak{F}(M) = \bigcap_{\varepsilon} \mathfrak{F}(M, \varepsilon)$. As the final piece of new notation we introduce $\mathfrak{E}(M)$ for the range of $E(M)$ whenever M is a Borel subset of the complex plane. In the next section we shall show that if M is compact, then $\mathfrak{F}(M) = \mathfrak{E}(M)$.

The point of our procedure is this. We are trying to prove that whenever A commutes with an operator B, then $E \leftrightarrow B$. For this purpose we need a direct, geometric characterization of $\mathfrak{E}(M)$ in terms of A, and that is exactly what the equation $\mathfrak{F}(M) = \mathfrak{E}(M)$ (whenever it is valid) gives us. Consideration of the subspaces $\mathfrak{F}(M)$ is quite natural when they are viewed from the proper value point of view mentioned in the preceding paragraph.

We conclude this section by borrowing the already announced result of the next section and, on the basis of that loan, proving the main commutativity theorem.

THEOREM 2. *If E is a compact, complex spectral measure and if B is an operator such that $\int \lambda dE(\lambda) \leftrightarrow B$, then $E \leftrightarrow B$.*

Proof. Using the notation established above, we begin by observing that, according to Theorem 1, $\mathfrak{F}(\lambda, \varepsilon)$ is invariant under B for all λ and ε. Since the span and the intersection of invariant subspaces are themselves invariant, it follows that the subspace $\mathfrak{F}(M)$ is invariant under B for every set M of complex numbers. If M is compact, then (by §42) we may conclude that $\mathfrak{E}(M)$ is invariant under B. The regularity of E shows then that $\mathfrak{E}(M)$ is invariant under B for every Borel set M. Since if M is a Borel set, then so is its complement M', and since $\mathfrak{E}(M') = (\mathfrak{E}(M))^{\perp}$, it follows that $\mathfrak{E}(M)$ reduces B for every Borel set M, and this is a paraphrase of what we have promised to prove.

§42. Characterization of the Spectral Subspaces

We continue to use the symbols E, A, $\mathfrak{E}$, and $\mathfrak{F}$, in the sense in which they were defined in the preceding section.

THEOREM 1. *If M is a Borel subset of the complex plane, then*

$$\mathfrak{E}(M) \subset \mathfrak{F}(M).$$

Proof. Let ε be a fixed positive number and let $\{M_j\}$ be a disjoint, countable family of non-empty Borel sets of diameter less than ε such that $\bigcup_j M_j = M$. For each index j, let λ_j be a complex number in M_j. If $x \in \mathfrak{E}(M)$ and if $x_j = E(M_j)x$, then

$$\| (A - \lambda_j)^n x_j \|^2 = \int | (\lambda - \lambda_j)^n |^2 d(E(\lambda)x_j, x_j)$$

for all j and n. Since $E(M_j')x_j = 0$ (where M_j' is the complement of M_j), it follows that

$$\| (A - \lambda_j)^n x_j \|^2 = \int_{M_j} | (\lambda - \lambda_j)^n |^2 d(E(\lambda)x_j, x_j) \leqq \varepsilon^{2n} \| x_j \|^2$$

for all j and n, and hence that $x_j \in \mathfrak{F}(\lambda_j, \varepsilon)$ for all j. Since

$$\mathfrak{F}(\lambda_j, \varepsilon) \subset \mathfrak{F}(M, \varepsilon)$$

and since $x = E(M)x = \Sigma_j E(M_j)x = \Sigma_j x_j$, it follows that $x \in \mathfrak{F}(M, \varepsilon)$. The arbitrariness of ε implies that $x \in \mathfrak{F}(M)$ and hence that

$$\mathfrak{E}(M) \subset \mathfrak{F}(M).$$

THEOREM 2. *If M is a compact subset of the complex plane, then* $\mathfrak{F}(M) \subset \mathfrak{E}(M)$.

Proof. Let M' be the complement of M, and let N be any compact subset of M'. If δ is the distance between the two compact sets M and N, then $\delta > 0$ and consequently we may find a number ε such that

$$0 < \varepsilon < \delta.$$

If $\lambda_0 \in M$ and if $x \in \mathfrak{F}(\lambda_0, \varepsilon)$, then $\| (A - \lambda_0)^n x \| \leqq \varepsilon^n \| x \|$ for every n; if on the other hand $x \in \mathfrak{E}(N)$, then

$$\| (A - \lambda_0)^n x \|^2 = \int_N | (\lambda - \lambda_0)^n |^2 d(E(\lambda)x, x) \geqq \delta^{2n} \| x \|^2.$$

It follows that no vector other than 0 can belong to both $\mathfrak{F}(\lambda_0, \varepsilon)$ and $\mathfrak{E}(N)$, i.e. that $\mathfrak{F}(\lambda_0, \varepsilon) \cap \mathfrak{E}(N) = \mathfrak{O}$. We propose to show that much more is true; we shall, in fact, prove that $\mathfrak{F}(\lambda_0, \varepsilon) \perp \mathfrak{E}(N)$. Since $E(N) \leftrightarrow A$, it follows from 41.1 that $\mathfrak{F}(\lambda_0, \varepsilon)$ is invariant under $E(N)$. Since a projection is a Hermitian operator, we may conclude that $\mathfrak{F}(\lambda_0, \varepsilon)$ reduces $E(N)$, or, equivalently, that the projection $F(\lambda_0, \varepsilon)$ with range $\mathfrak{F}(\lambda_0, \varepsilon)$ commutes with $E(N)$. We know therefore that the product $F(\lambda_0, \varepsilon)E(N)$ is the projection with range $\mathfrak{F}(\lambda_0, \varepsilon) \cap \mathfrak{E}(N)$, and hence that $F(\lambda_0, \varepsilon)E(N) = 0$. This, however, is what we promised: $\mathfrak{F}(\lambda_0, \varepsilon)$ is indeed orthogonal to $\mathfrak{E}(N)$. The arbitrariness of λ_0 in M implies that $\mathfrak{F}(M, \varepsilon) \perp \mathfrak{E}(N)$ and hence that $\mathfrak{F}(M) \perp \mathfrak{E}(N)$. To sum up: if N is a compact subset of M', then $\mathfrak{F}(M) \perp \mathfrak{E}(N)$. The regularity of E implies that $\mathfrak{F}(M) \perp \mathfrak{E}(M') = (\mathfrak{E}(M))^{\perp}$, and hence that

$$\mathfrak{F}(M) \subset \mathfrak{E}(M),$$

as asserted.

§43. The Spectral Theorem for Hermitian Operators

It is high time to prove that in the course of the last several sections we have not been operating in a vacuum. The following theorem settles all such doubts for Hermitian operators.

THEOREM 1. *If A is a Hermitian operator, then there exists a (neces-*

sarily real and necessarily unique) compact, complex spectral measure E, called the spectral measure of A, such that $A = \int \lambda \, dE(\lambda)$.

Proof. Let x and y be any two fixed vectors and write

$$L(p) = (p(A)x, y)$$

for every real polynomial p. It follows from 34.3 that

$$| L(p) | \leqq \mathbf{N}_A(p) \cdot || x || \cdot || y ||$$

and hence that, with respect to the norm $\mathbf{N}_A$, L is a bounded linear functional of its argument. There exists consequently a unique complex measure μ in the compact set $\Lambda(A)$ such that $(p(A)x, y) = \int p(\lambda) \, d\mu(\lambda)$ for every real polynomial p and such that $| \mu(M) | \leqq || x || \cdot || y ||$ for every Borel set M. We shall find it convenient to indicate the dependence of μ on x and y by writing $\mu_M(x, y)$ instead of $\mu(M)$.

Using the uniqueness of μ, we may proceed by straightforward computations to prove that μ_M is a symmetric, bilinear functional for each Borel set M. The proof of the fact that μ_M is additive in its first argument runs, for instance, as follows:

$$\int p(\lambda) \, d\mu_\lambda(x_1 + x_2, y) = (p(A)(x_1 + x_2), y) = (p(A)x_1, y) + (p(A)x_2, y)$$
$$= \int p(\lambda) \, d\mu_\lambda(x_1, y) + \int p(\lambda) \, d\mu_\lambda(x_2, y).$$

Since, in virtue of the relation $| \mu_M(x, y) | \leqq || x || \cdot || y ||$, valid for all M, x, and y, the bilinear functionals μ_M are bounded, it follows that for each M there exists a unique Hermitian operator $E(M)$ such that $\mu_M(x, y) = (E(M)x, y)$ for all x and y. Consideration of the polynomials p_0 and p_1, defined by $p_0(\lambda) = 1$ and $p_1(\lambda) = \lambda$, implies that

$$\int d(E(\lambda)x, y) = (E(X)x, y) = (x, y)$$

and

$$\int \lambda d(E(\lambda)x, y) = (Ax, y)$$

for all x and y. In view of 36.3, all that remains in order to complete the proof of the theorem is to establish that the function E is projection-valued; we shall do this by proving that E is multiplicative.

For any fixed pair of vectors x and y and for any real polynomial q, we introduce the auxiliary complex measure ν defined for every Borel set M by $\nu(M) = \int_M q(\lambda) \, d(E(\lambda)x, y)$. If p is any real polynomial, then

$$\int p(\lambda) d\nu(\lambda) = \int p(\lambda) q(\lambda) \, d(E(\lambda)x, y) = (p(A)q(A)x, y)$$
$$= (p(A)x, q(A)y) = \int p(\lambda) \, d(E(\lambda)x, q(A)y)$$

and therefore

$$\nu(M) = \int q(\lambda)\chi_M(\lambda)\, d(E(\lambda)x, y) = (E(M)x, q(A)y)$$
$$= (q(A)E(M)x, y) = \int q(\lambda)\, d(E(\lambda)E(M)x, y)$$

for every Borel set M. Since q is arbitrary, it follows that

$$(E(M \cap N)x, y) = \int_{M\cap N} d(E(\lambda)x, y)$$
$$= \int_N \chi_M(\lambda)\, d(E(\lambda)x, y) = (E(N)E(M)x, y)$$

for every Borel set N, and hence that $E(M \cap N) = E(M)E(N)$. The proof of the spectral theorem for Hermitian operators is thereby complete.

Although our proof of this theorem appears at a rather late stage of the development of the theory, the proof does not, as a matter of fact, use much of that theory. In addition to the very elements of Hilbert space geometry, and the external analytic crutch of measure theory, the proof relies on the connection between bilinear functionals and operators and on the connection between the norm and the spectrum of a Hermitian operator. We recall that the first of these connections is based on the representation of linear functionals by vectors, and that the second one (which is the one that really exploits the Hermitian character of A) involves the elementary properties of the concepts of spectrum and approximate point spectrum. Almost none of the information that we have accumulated about spectral measures was needed, and only superficial (but apparently unavoidable) use was made of the fact that A is Hermitian; we did not even need to know the slightly tricky relation $\| A \| = \sup \{| (Ax, x) |: \| x \| = 1\}$. The proof applies, of course, to the special case in which $\mathfrak{H}$ is finite-dimensional. In view of the lot of apparently formidable machinery that we have used, this last comment might appear silly—the spectral theorem for the finite-dimensional case is, after all, quite near the surface. A closer examination of the facts shows however that, since the measure-theoretic apparatus becomes almost vacuous in the finite case, our procedure yields a rather reasonable proof even there. The reader who is not quite clear as to exactly which concepts are needed exactly where would do well to retrace our steps and examine the extent to which they become simplified in the presence of finite dimensionality.

§44. The Spectral Theorem for Normal Operators

THEOREM 1. *If A is a normal operator, then there exists a (necessarily unique) compact, complex spectral measure E, called the spectral measure of A, such that $A = \int \lambda dE(\lambda)$.*

Proof. If A_1 and A_2 are the real and imaginary parts of A respectively, i.e. if A_1 and A_2 are Hermitian operators such that

$$A = A_1 + iA_2,$$

then, by the theorem of the preceding section, there exist two real, compact spectral measures E_1 and E_2 such that

$$A_1 = \int \lambda \, dE_1(\lambda)$$

and

$$A_2 = \int \lambda \, dE_2(\lambda).$$

It will be convenient to regard the complex plane as the Cartesian product of the real and the imaginary axes. In accordance with this view, we shall use the term *rectangle* to stand for the Cartesian product $M_1 \times M_2$ of a Borel subset M_1 of the real axis and a Borel subset M_2 of the imaginary axis. Since the fact that A is normal implies that all operators in sight (and in particular $E_1(M_1)$ and $E_2(M_2)$) commute with each other, it follows that $E_1(M_1) E_2(M_2)$ is a projection. The remainder of our discussion will be devoted to sketching the proof of the fact that there exists a (necessarily compact) complex spectral measure E such that if $M = M_1 \times M_2$ is a rectangle, then $E(M) = E_1(M_1) E_2(M_2)$. We leave it to the reader to verify that a spectral measure E with this property also has the property that $\int \lambda \, dE(\lambda) = A_1 + iA_2 = A$; the verification depends on the fact that if a function on a product space is independent of one of its two possible arguments, then its integral can be evaluated by an integration on the other one of the two factor spaces.

For any fixed vector x let $\hat{\mu}$ be the function of rectangles defined by $\hat{\mu}(M_1 \times M_2) = (E_1(M_1)E_2(M_2)x, x)$. The properties of the spectral measures E_1 and E_2 imply that $\hat{\mu}$ is non-negative, finitely additive, and continuous from below in the sense that its value on the union of an increasing sequence of rectangles is the limit of its values on the terms of the sequence. It follows that $\hat{\mu}$ can be extended to a measure on the class of all Borel sets in the complex plane. It is convenient to indicate the dependence of the extended $\hat{\mu}$ on x by denoting its value on any Borel set M by $\hat{\mu}_M(x)$.

For every Borel set M and for every pair of vectors x and y we write

$$\mu_M(x, y) = \hat{\mu}_M(\tfrac{1}{2}(x + y)) - \hat{\mu}_M(\tfrac{1}{2}(x - y)) + i\hat{\mu}_M(\tfrac{1}{2}(x + iy)) - i\hat{\mu}_M(\tfrac{1}{2}(x - iy)).$$

We assert that μ_M is, for each fixed Borel set M, a symmetric bilinear functional. This assertion is proved by noting that (i) it is true if M

is a rectangle, and (ii) the class of all sets for which it is true is closed under the formation of complements and countable unions. Since

$$| \mu_M(x, x) | = | \hat{\mu}_M(x) | \leqq \| x \|^2$$

whenever M is a rectangle, it follows that, for each Borel set M, the bilinear functional μ_M is bounded (and has, in fact, a norm not exceeding 1). We are almost at the end: by now we know that to every Borel set M there corresponds a bounded Hermitian operator $E(M)$ and that the function E has all the required properties except possibly multiplicativity.

The last point is settled as follows. Fix x and y and, for each pair of Borel sets M and N, consider the two expressions $(E(M \cap N)x, y)$ and $(E(M)E(N)x, y)$. If N is a rectangle, then the class of all sets M for which these two expressions are equal is such that (cf. the preceding paragraph) (i) it contains all rectangles and (ii) it is a Boolean σ-algebra. Consequently this class contains all Borel sets. The same argument may now be applied to prove that, for each fixed Borel set M, the class of admissible N's is also equal to the class of all Borel sets, and thus the proof grinds to a stop.

CHAPTER III

THE ANALYSIS OF SPECTRAL MEASURES

§45. The Problem of Unitary Equivalence

Now where are we? The main purpose of the study of operator theory is to discover, formulate, and prove the proper generalizations, valid for all Hilbert spaces, of the powerful results known in the finite-dimensional case. In so far as these results concern normal operators they are all easy consequences of the possibility of reducing normal matrices to diagonal form. The diagonalization theorem yields, in particular, the ultimate desideratum, namely a complete description of the geometric behaviour of all normal matrices. Speaking slightly more explicitly we may say that the diagonalization theorem gives us a method which enables us to construct all possible normal operators on a finite-dimensional Hilbert space. The construction is based on such elementary and completely manageable material as the concept of a finite set of complex numbers. Although the general spectral theorem for normal operators is frequently asserted to be the infinite-dimensional analog of diagonalization, it is nowhere near as powerful as its purely algebraic special case. The spectral theorem does not, for instance, tell us how to construct all possible normal operators. All that the spectral theorem does accomplish in this direction is to reduce the problem to the construction of all possible spectral measures, and thereby, probably, to leave the prospective constructor more bewildered than he was at the beginning.

These remarks are offered by way of introduction to the circle of ideas usually called the problem of unitary equivalence. Two operators A and B are *equivalent* if there is an automorphism U of the underlying Hilbert space $\mathfrak{H}$ which carries A onto B, or, in more detail, if there exists a unitary operator U such that $U^{-1}AU = B$. The problem of unitary equivalence is to find necessary and sufficient conditions on A and B for the existence of such a U. Since equivalent operators are geometrically indistinguishable, any "description" of an operator A will at the same time be a description of all operators belonging to the same equivalence class as A. In other words: since, geometrically speaking, A is only determined to within unitary equivalence, a more delicate description of A neither should nor does exist.

If A is a normal operator with spectral measure E, the problem of finding all operators equivalent to A is settled, in a certain repulsive sense, by the equation $U^{-1}AU = \int \lambda\, d(U^{-1}E(\lambda)U)$. (The symbol U in this equation denotes, of course, a unitary operator; the interpretation and proof of the equation are achieved by the formation of the usual inner products and should be obvious to the reader who has followed the development of spectral theory so far.) If, in other words, we say that two spectral measures E and F, with the same domain, are *equivalent* whenever there exists a unitary operator U such that

$$U^{-1}E(M)U = F(M)$$

for all M in the common domain of definition of E and F, then a necessary and sufficient condition for the equivalence of two normal operators is the equivalence of their respective spectral measures.

The main reason for feeling dissatisfied with the above answer to the equivalence problem is that it leaves things pretty much where they were: in order to decide whether or not two given operators are equivalent we must still ask, separately for each unitary operator, whether or not it is willing to perform the miracle required of it. What is really wanted is a *complete set of invariants* for the unitary equivalence of normal operators. In qualitative terms this means that we wish to associate with each normal operator A a certain "object" u_A so that the following conditions are satisfied. (i) If A and B are equivalent normal operators, then $u_A = u_B$. (ii) If A and B are normal operators such that $u_A = u_B$, then A and B are equivalent. (iii) To every object u there corresponds at least one normal operator A such that $u_A = u$. (iv) The objects u are easily manageable mathematical concepts, which may be described in simple and, as far as possible, constructive terms, and whose definition is, preferably, independent of operator theory. It is worth while to note in passing that the spectrum $\Lambda(A)$ of a (not even necessarily normal) operator A satisfies conditions (i), (iii), and (iv). We may therefore say that the points of $\Lambda(A)$ constitute a set of unitary invariants of A, but not a complete set of such invariants.

The first three of the above conditions describe nothing more than a one-to-one mapping from the set of all equivalence classes of normal operators onto the set of all objects. The reader who reads on to finish this book will, at the end, be in a position to judge whether or not our solution satisfies the last condition. In order to motivate both the result we shall obtain and the method we shall use to get it, we begin, in the next section, by taking a closer look at the situation in finite-dimensional spaces.

§46. Multiplicity Functions in Finite-dimensional Spaces

The set of all proper values of a normal operator A on a finite-dimensional Hilbert space $\mathfrak{H}$, together with their associated multiplicities, form a complete set of unitary invariants of A. These invariants may be described as follows. To the normal operator A there corresponds a function $u = u_A$; the domain of u is the complex plane and the values of u are finite cardinal numbers. (The value $u(\lambda)$ of the function u at the complex number λ is to be interpreted as the multiplicity with which λ occurs as a proper value of A; if λ is not a proper value of A at all, we write $u(\lambda) = 0$.) Not every function with the indicated domain and range arises in this manner from some normal operator A. In order that a function u do come from some A it is, in fact, necessary and sufficient that the sum of all the values of u be the dimension of the Hilbert space $\mathfrak{H}$ (and hence, in particular, it is necessary that u vanish at all but a finite number of points). Anyone familiar with the diagonalization theory of normal matrices can verify at a glance that the function u_A satisfies all the conditions stated and discussed in the preceding section.

To prepare the way for understanding the generalized version of multiplicity functions such as u_A, we proceed to describe them in different terms. Since infinite-dimensional spectral measures associate projections with Borel sets of complex numbers, and not with individual complex numbers, it ought not to be surprising that we get a nearer approximation to the final version of multiplicity theory if we regard the domain of a multiplicity function as the class of all Borel sets in the complex plane, and not as the complex plane itself. The transition in point of view is easy: for any non-empty Borel set M we define $u_A(M)$ to be the minimum value of $u_A(\lambda)$ for all λ in M; for $M = 0$ we write $u_A(M) = 0$.

Not every function u whose domain is the class of all Borel sets in the complex plane and whose values are finite cardinal numbers is the multiplicity function of some normal operator A on a finite-dimensional Hilbert space. It is easy to verify that if u does come from some A, then

$$u(M) \geqq u(N)$$

whenever M and N are Borel sets such that $0 \neq M \subset N$, and

$$u(\bigcup_n M_n) = \min_n \{u(M_n)\}$$

whenever $\{M_n\}$ is a disjoint sequence of Borel sets. Even these conditions are only necessary; they are not yet sufficient to ensure the existence of an A such that $u = u_A$. It is easily possible to adjoin to these conditions a finiteness requirement such that together with it they become necessary and sufficient. Since, however, the conditions already stated are the only ones that persist in the general (not necessarily finite-dimensional) case, we shall not bother to formulate the extra one that applies only provincially.

Unfortunately we are still far from the definition of the kind of multiplicity function that really arises in infinite-dimensional cases. The difficulty is that the concept of a set (Borel set or not) is not quite the relevant one. The argument of our general multiplicity function will not be a set but a finite measure. Speaking very roughly a finite measure μ in the complex plane may be considered as a set. What we have in mind is "the set on which μ is concentrated" or "the complement of the largest set on which μ vanishes." Such phrases are nonsense of course. It is, however, true that a measure μ for which there exists a finite set on whose complement μ vanishes is in an obvious sense a generalization of a finite set. Enough of the sense in which this is true carries over to the infinite case that a successful theory can be built on it. We must, however, postpone further discussion of these matters until after the presentation of the pertinent properties of measures.

§47. Measures

Let $(X, \mathbf{S})$ be a measurable space; the only measures that we shall consider from now on are finite measures whose domain of definition is $\mathbf{S}$. We recall that a measure ν is *absolutely continuous* with respect to a measure μ, in symbols $\nu \ll \mu$, if $\nu(M) = 0$ for every set M such that $\mu(M) = 0$. We shall have occasion to use the *Radon-Nikodym theorem;* it asserts that if μ and ν are measures such that $\nu \ll \mu$, then there exists a non-negative function f in $\mathfrak{L}_1(\mu)$ such that $\nu(M) = \int_M f\, d\mu$ for every set M in $\mathbf{S}$. A measure μ is *equivalent* to a measure ν, in symbols $\mu \equiv \nu$, if $\mu \ll \nu$ and $\nu \ll \mu$; it is obvious that the terminology is justified, or, in other words, that the relation $\equiv$ is an equivalence.

If μ is a measure and $M \in \mathbf{S}$, we shall write μ_M for the measure defined for every N in $\mathbf{S}$ by $\mu_M(N) = \mu(M \cap N)$.

THEOREM 1. *If μ is a measure and if M and N are in $\mathbf{S}$, then a necessary and sufficient condition that $\mu_M \ll \mu_N$ is that $\mu(M - N) = 0$.*

Proof. If $\mu(M - N) = 0$ and if M_0 is a set in $\mathbf{S}$ such that $\mu_N(M_0) = 0$, then

$$\mu_M(M_0) = \mu(M \cap M_0) = \mu((M \cap N) \cap M_0) + \mu((M - N) \cap M_0) = 0.$$

If, conversely, $\mu_M \ll \mu_N$, then, since we have $\mu_N(M - N) = 0$, it follows that $\mu(M - N) = \mu_M(M - N) = 0$.

THEOREM 2. *If μ and ν are measures such that $\nu \ll \mu$, then there exists a set N in $\mathbf{S}$ such that $\nu \equiv \mu_N$.*

Proof. By the Radon-Nikodym theorem there exists a non-negative function f in $\mathfrak{L}_1(\mu)$ such that $\nu(M) = \int_M f(t)\, d\mu(t)$ whenever $M \in \mathbf{S}$. If $N = \{t : f(t) > 0\}$, then $\int_{M-N} f\, d\mu = 0$ and therefore

$$\nu(M) = \int_{M \cap N} f\, d\mu$$

whenever $M \in \mathbf{S}$. It follows that $\nu(M) = 0$ if and only if $\mu(N \cap M) = 0$, i.e. that $\nu \equiv \mu_N$.

The objects of principal interest for us will be not measures but equivalence classes of measures. In order, however, to minimize complications, we shall adopt a point of view similar to that frequently adopted in number theory. (It is easier to discuss integers and congruence than to discuss equivalence classes of integers and equality.) We shall accordingly formulate definitions and announce theorems about measures, intending all the while that our statements should be interpreted so as to apply to equivalence classes of measures. An alternative point of view is to think of a measure as the class of all sets on which it vanishes. The intuitively most helpful attitude is to think of a measure as being the same as "the" set on which it is concentrated; cf. Theorems 1 and 2 and the remarks at the end of the preceding section. In order to minimize the possibility of confusion we shall, however, continue to distinguish by our notation between equality in fact ($\mu = \nu$) and equality by convention ($\mu \equiv \nu$).

A typical statement which must be interpreted in terms of equivalence is that with respect to ordering by absolute continuity the set of all measures is a partially ordered set. In technical language (which we shall in fact not employ) our result will be that the partially ordered set of all measures is a Boolean σ-ring with the property that every principal ideal satisfies the countable chain condition.

§48. Boolean Operations on Measures

We continue to use the notations and conventions of the preceding section and, in particular, the use of the word "measure" to mean "finite measure." If μ_1 and μ_2 are measures, then there exists a measure μ such that $\mu_1 \ll \mu$, $\mu_2 \ll \mu$, and such that $\mu \ll \nu$ whenever the measure ν is such that $\mu_1 \ll \nu$ and $\mu_2 \ll \nu$. In other words, the supremum

$$\mu \equiv \mu_1 \vee \mu_2$$

of any two elements μ_1 and μ_2 of the partially ordered set of all measures is another element of that set; the proof of this assertion is achieved simply by writing $\mu = \mu_1 + \mu_2$. With a very small modification the same technique may be used to show the existence of the supremum $\bigvee_j \mu_j$ of any countable family of measures. There is, indeed, no loss of generality in assuming that $\Sigma_j \mu_j(X) < \infty$—if this were not already true, we could make it true by, for instance, replacing μ_j by a suitable, small positive multiple of μ_j. (The replacement yields a measure equivalent to μ_j.) It follows that the set function μ defined for each M in $\mathbf{S}$ by $\mu(M) = \Sigma_j \mu_j(M)$ is a measure; it is clear that $\mu \equiv \bigvee_j \mu_j$. In view of Theorem 1 below, it is not even necessary to verify the last assertion; all that is needed from our discussion is the fact that every countable family $\{\mu_j\}$ of measures is bounded. (To say that a family $\{\mu_j\}$ of measures is *bounded* means that there exists a measure μ such that $\mu_j \ll \mu$ for all values of j.)

THEOREM 1. *Every bounded family* $\{\mu_j\}$ *of measures has a supremum and, in fact,* $\{\mu_j\}$ *has a countable subfamily* $\{\mu_{k_n}\}$ *such that* $\bigvee_j \mu_j \equiv \bigvee_n \mu_{k_n}$.

Proof. Let μ be a measure such that $\mu_j \ll \mu$ for every j, and, for each j, let N_j be a set in $\mathbf{S}$ such that $\mu_j \equiv \mu_{N_j}$. Form all finite unions of the N_j's, evaluate μ on each such union, and let α be the supremum of the numbers so obtained. If $\{M_n\}$ is a sequence of such finite unions with the property that $\mu(M_n) \to \alpha$, and if $M = \bigcup_n M_n$, then $\mu_M \equiv \bigvee_j \mu_j$. Indeed if $\mu(N_j - M) > 0$ for some j, then $\mu(N_j \cup M_n) > \alpha$ for some n and, since this contradicts the definition of α, we have $\mu_j \ll \mu_M$. If, on the other hand, ν is a measure such that $\mu_j \ll \nu$ for every j, then

$\mu_{N_j} \ll \nu$ for every j. It follows that $\mu_{M_n} \ll \nu$ for every n and hence that $\mu_M \ll \nu$.

It follows from Theorem 1 that every non-empty family $\{\mu_j\}$ of measures has an infimum, $\bigwedge_j \mu_j$; the infimum is obtained by forming the supremum of the family of all measures bounded by every μ_j. Consequently the partially ordered set of all measures is not only a lattice, but even a σ-lattice, and a boundedly complete lattice. An application of 47.2, similar to the one made in the proof of Theorem 1, shows that this lattice is distributive. The main point of 47.2 is exactly its applicability to such situations; by means of it most of the algebraic facts concerning measures may be reduced to the corresponding algebraic facts concerning sets.

It is convenient to say that two measures μ and ν are *orthogonal*, in symbols $\mu \perp \nu$, if $\mu \wedge \nu = 0$; a family $\{\mu_j\}$ of measures is an *orthogonal family* if $\mu_j \perp \mu_k$ whenever $j \neq k$. Another example of the sort of application of 47.2 that was mentioned in the preceding paragraph occurs in the proof of the assertion that a bounded orthogonal family of non-zero measures is necessarily countable.

Our next and last result about the algebra of measures asserts that the set of all measures is not only a distributive lattice but is in fact quite anxious to look and act like a Boolean algebra. There is in general no "unit" measure, i.e. the set of *all* measures is not in general bounded, and it is therefore not only false but even meaningless to say that every measure has a complement. It does, however, make sense to speak of relative complements, or differences, and that is what Theorem 2 does.

THEOREM 2. *If μ and ν are measures, then there exists a measure μ_0 such that $\mu_0 \perp \nu$ and $\mu_0 \vee \nu \equiv \mu \vee \nu$.*

Proof. If $\nu \ll \mu$, then $\nu \equiv \mu_N$ for some N in $\mathbf{S}$, and μ_{X-N} does everything expected of μ_0. In the general case (i.e. when ν is not necessarily bounded by μ) this special result may be applied to $\mu \vee \nu$ and ν in place of μ and ν respectively.

§49. Multiplicity Functions

We are now in a position to describe the objects which will occur as complete sets of unitary invariants for spectral measures. A *multiplicity function* is a function u whose values are (not necessarily finite) cardinal numbers, whose domain is the set of all finite measures in a measurable space $(X, \mathbf{S})$, and which satisfies the following three conditions: (i) if μ is the measure which is identically zero, then $u(\mu) = 0$; (ii) if μ and ν are measures such that $0 \neq \nu \ll \mu$, then $u(\mu) \leqq u(\nu)$; and (iii) if a measure μ is the supremum of a countable orthogonal family $\{\mu_j\}$ of non-zero measures, then $u(\mu) = \min \{u(\mu_j)\}$. We observe that since a bounded orthogonal family of non-zero finite measures is necessarily countable, the third condition is only vacuously strengthened by removing from it the word "countable."

It is not hard to give examples of multiplicity functions. Given the measurable space $(X, \mathbf{S})$, let $\{\mu_j\}$ be an arbitrary orthogonal family of finite, non-zero measures on $\mathbf{S}$ and, for each j, let u_j be a cardinal number. If a non-zero measure μ is covered by the family $\{\mu_j\}$ in the sense that $\mu \equiv \bigvee_j(\mu \wedge \mu_j)$, we define $u(\mu)$ to be the smallest one of those cardinal numbers u_j for which $\mu \wedge \mu_j \neq 0$; for all other measures μ we define $u(\mu)$ to be 0. We leave to the reader the verification that the u so defined is indeed a multiplicity function, and we turn to the more important task of proving that every multiplicity function may be obtained in this manner.

To motivate our procedure we take one more look at the example of the preceding paragraph. If j and k are indices such that $u_j < u_k$, then $u(\mu_j \vee \mu_k) = u(\mu_j) = u_j$. It is really possible, in other words, that the second condition in the definition of multiplicity functions is not vacuously satisfied, i.e. that μ and ν are measures such that $0 \neq \nu \ll \mu$ and $u(\mu) < u(\nu)$. It is natural to say that if for a given measure μ this never happens, if, that is, $u(\mu) = u(\nu)$ whenever $0 \neq \nu \ll \mu$, then μ has *uniform multiplicity*. In the example of the preceding paragraph this concept is illustrated by each term of the defining family; it is true, in other words, that μ_j has uniform multiplicity (equal to u_j) for each value of j.

THEOREM 1. *If u is a multiplicity function and if μ is a non-zero finite measure on $\mathbf{S}$, then there exists a non-zero measure μ_0 such that $\mu_0 \ll \mu$ and such that μ_0 has uniform multiplicity.*

Proof. Write $\bar{\nu} \equiv \bigvee \{\nu : \nu \ll \mu,\ u(\nu) > u(\mu)\}$, and (48.2) let μ_0 be a measure such that $\mu_0 \perp \bar{\nu}$ and $\mu_0 \vee \bar{\nu} \equiv \mu$. Since (48.1) $\bar{\nu}$ may also be expressed as the supremum of a countable family of measures ν for which $u(\nu) > u(\mu)$, and since a standard use of 48.2 shows that this countable family may be assumed to be orthogonal, it follows that $u(\bar{\nu}) > u(\mu)$. Since $\mu \equiv \mu_0 \vee \bar{\nu}$, it follows that $\mu_0 \neq 0$ and hence that

$u(\mu_0) = u(\mu)$. To prove that μ_0 has uniform multiplicity, suppose that $0 \neq \nu \ll \mu_0$. Since the assumption $u(\nu) > u(\mu_0) = u(\mu)$ leads to the contradiction $\nu \ll \bar{\nu}$, the proof is complete.

THEOREM 2. *If u is a multiplicity function and if μ is a non-zero finite measure on* S, *then there exists a (necessarily countable) orthogonal family $\{\mu_j\}$ of non-zero measures such that each μ_j has uniform multiplicity and such that $\mu \equiv \vee_j \mu_j$.*

Proof. In virtue of Theorem 1 there do exist orthogonal families of non-zero measures each term of which is bounded by μ and has uniform multiplicity; let $\{\mu_j\}$ be a maximal family with these properties. If $\vee_j \mu_j \equiv \nu$ and if $\nu \not\equiv \mu$, then, by 48.2, there exists a non-zero measure bounded by μ and orthogonal to ν. An application of Theorem 1 to that measure shows that its existence contradicts the maximality of the family $\{\mu_j\}$ and it follows that $\vee_j \mu_j \equiv \mu$.

THEOREM 3. *If u is a multiplicity function, then there exists an orthogonal family $\{\mu_j\}$ of non-zero finite measures on* S *such that each μ_j has uniform multiplicity and such that $\mu \equiv \vee_j(\mu \wedge \mu_j)$ whenever μ is a finite measure on* S.

Proof. Select a maximal orthogonal family of non-zero finite measures on S and apply Theorem 2 to each term of that family. We may collect the resulting family of families into one family $\{\mu_j\}$ which will then be a maximal orthogonal family of non-zero measures and which will, in addition, have the property that each μ_j has uniform multiplicity. It remains merely to prove that if $\{\mu_j\}$ is a maximal orthogonal family of non-zero finite measures on S, then $\mu \equiv \vee_j(\mu \wedge \mu_j)$ for every finite measure μ on S. The argument for this purpose proceeds just as in the proof of Theorem 2. If, for a given μ, $\vee_j(\mu \wedge \mu_j) \equiv \nu$, and if $\mu \not\equiv \nu$, then, by 48.2, there exists a measure μ_0 such that $0 \neq \mu_0 \ll \mu$ and $\mu_0 \perp \nu$. Since it follows that $\mu_0 \wedge \mu_j \equiv \mu_0 \wedge (\mu \wedge \mu_j) \ll \mu_0 \wedge \nu = 0$, i.e. that $\mu_0 \perp \mu_j$ for all j, this contradicts the maximality of the family $\{\mu_j\}$ and proves therefore the relation $\mu \equiv \vee_j(\mu \wedge \mu_j)$.

It is clear that Theorem 3 implies what we promised to show, i.e. that every multiplicity function may be obtained in the way in which we obtained our first example. We cannot, of course, assert that the family $\{\mu_j\}$ described in Theorem 3 is uniquely determined by the multiplicity function u; several applications of Zorn's lemma have cut us off from being able to claim any naturality for the objects whose existence we proved.

§50. The Canonical Example of a Spectral Measure

Suppose that $(X, \mathbf{S})$ is a measurable space, $\{\mu_j\}$ is an orthogonal family of non-zero finite measures on $\mathbf{S}$, and, for each value of j, u_j is a cardinal number. For each value of j we consider the Hilbert space obtained by forming the direct sum of u_j copies of $\mathfrak{L}_2(\mu_j)$ and we (temporarily) denote by $\mathfrak{H}$ the direct sum (over the index j) of the Hilbert spaces so obtained. A typical element of $\mathfrak{H}$ is a doubly indexed family $\{f_{jk}\}$ of functions on X such that $f_{jk} \in \mathfrak{L}_2(\mu_j)$ for each j and k; for a fixed value of j the index k has u_j possible values. By the *canonical spectral measure* associated with the families $\{\mu_j\}$ and $\{u_j\}$ we mean the spectral measure E defined for each M in $\mathbf{S}$ by $E(M)\{f_{jk}\} = \{\chi_M f_{jk}\}$.

One of our results will be that upon the application of a suitable isomorphism every spectral measure may be put into this canonical form. Applying that result to compact, complex spectral measures we conclude that every normal operator is isomorphic to a direct sum of multiplications by bounded measurable functions on finite measure spaces, or, equivalently, that it is isomorphic to a multiplication by a bounded measurable function on a direct sum of finite measure spaces. (We have not given and we need not and will not give the detailed definition of the latter concept.) Another way of expressing this result is to say that a suitable (in general highly infinite) measure μ may be introduced into the spectrum of any normal operator A so that A becomes isomorphic to the multiplication operator which sends each function f in $\mathfrak{L}_2(\mu)$ on the function g defined by $g(\lambda) = \lambda f(\lambda)$. Since all these statements will be immediate consequences of our study of spectral measures, we shall devote our attention to spectral measures exclusively.

In terms of spectral measures it is easy to describe our intentions. We shall associate a multiplicity function u with every spectral measure E in such a way that if $\{\mu_j\}$ is any orthogonal family with the properties described in 49.3, then E is isomorphic to the canonical spectral measure associated with $\{\mu_j\}$ and $\{u(\mu_j)\}$. (Observe that this implies in particular that, despite the non-uniqueness of $\{\mu_j\}$, the canonical spectral measure is determined by u uniquely to within unitary equivalence.) It will follow that two spectral measures are equivalent if and only if they have the same multiplicity function, and consequently the proof of this result will indeed fulfill all our promises.

Let us now return to the canonical example described above. If $\mathfrak{K}$ is any one of the doubly indexed family of Hilbert spaces used to form $\mathfrak{H}$, then $\mathfrak{K}$ may be viewed as a subspace of $\mathfrak{H}$. Since the subspace $\mathfrak{K}$ is invariant under $E(M)$, for every set M in **S**, the projection P with range $\mathfrak{R}$ commutes with the spectral measure E. (If E is the spectral measure of a normal operator A, the last assertion may be reformulated by saying either that the subspace $\mathfrak{R}$ reduces A or that the projection P commutes with A.) These comments indicate that the projections P which commute with a spectral measure E are the building blocks out of which E is constructed and that the analysis of spectral measures ought, therefore, to analyze all such projections. In the next two sections we indicate the details of such an analysis in the finite-dimensional case; after that we shall finally be ready to enter with understanding into the technical details of the general case.

§51. Finite-dimensional Spectral Measures

Let E be the spectral measure of a normal operator A on a finite-dimensional Hilbert space $\mathfrak{H}$. Let $\{\lambda_j\}$ be the family of all distinct proper values of A; for each j, let E_j be the value of E on the set containing λ_j alone, and let u_j be the dimension of the range of E_j (i.e. the multiplicity of the proper value λ_j). It is in many respects helpful to consider a structure analogous to the one formed by the λ's, E's, and u's. The analogs of the λ's are to be points spaced at, say, unit distances apart on a horizontal line segment. The role of E_j is to be played by a finite set, corresponding to the base point λ_j and thought of as arranged in a vertical column standing over λ_j; it will be convenient to space the points of such a column so that each of them is at a unit distance from its nearest neighbors. The fact, finally, that u_j is the dimension number corresponding to E_j is to be indicated by letting the set corresponding to E_j have cardinal number u_j. The entire set-theoretic configuration thereby described is exemplified by the diagram below. If P is a projection which commutes with E, then the range of P is a subspace which

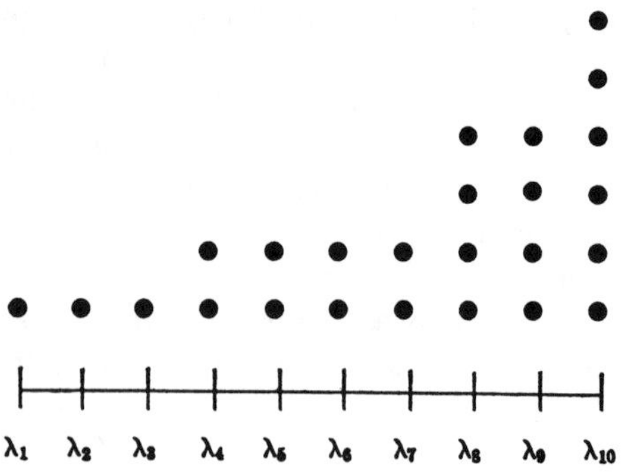

reduces A. The operator A when restricted to the range of P has its proper values among the λ_j's and is such that the proper space corresponding to each λ_j is a subspace of the range of E_j. The set-theoretic analog of a projection such as P is, therefore, a set obtained by selecting a (not necessarily proper and not necessarily non-empty) subset from each column and forming the union of the selected sets; in other words the analog of P is an arbitrary subset of the union of all columns. A distinguished role is played by the subsets which consist of entire columns; they are the analogs of the values of the spectral measure.

In accordance with our indications in and since §46, we shall think of multiplicity as defined not for proper values only but also for sets of proper values, or, equivalently, for arbitrary values of the spectral measure. If, for instance, the spectral measure E is such that its associated column configuration is exactly the one indicated by our diagram, then the multiplicity of the value of E on the entire complex plane is 1, and the multiplicity of $E(\{\lambda_7, \lambda_8, \lambda_9, \lambda_{10}\})$ is 2.

Once the diagram corresponding to a spectral measure has been constructed, it is trivial to read off from it the answer to every multiplicity question. The multiplicity associated with any set of λ's is the largest number of rows each of which cuts across the entire set under consideration. If the spectral measure is such that every column is of height 1 (if, in other words, every proper value is simple), then the answer to every multiplicity question is 1 or 0. Since the answer to the most general multiplicity question can be formulated in terms of rows, in terms, that is, of what may well be called simple spectral measures, it behooves us to try to understand the concept of simplicity and the manner in which a general spectral measure is made up of simple pieces.

§52. Simple Finite-dimensional Spectral Measures

The finite-dimensional case and the general case described in §50 make contact with each other through the following comment. Suppose that the finite-dimensional spectral measure E discussed in the preceding section is simple, i.e. that each u_j is equal to 1. Consider in this case the measurable space X whose points are the proper values of the operator A and all of whose subsets are measurable; let μ be the measure in X whose value on any subset of X is the number of points in that subset. It is easy to verify that, under these circumstances, the canonical spectral measure associated with μ (i.e. the one whose value on a set M is multiplication by the characteristic function of M) is isomorphic to E. In other words: the building blocks which served to construct our general canonical examples are natural generalizations of the simple pieces that occur in all finite-dimensional spectral measures.

The simple pieces may also be characterized intrinsically, without the use of an auxiliary measure space. Every vector x in our finite-dimensional Hilbert space $\mathfrak{H}$ may be written as a sum, $x = \Sigma_j x_j$, where, for each j, x_j belongs to the range of E_j. If we apply all possible values of E to x and then form the projection on the subspace spanned by the vectors so obtained, we end up with a projection P such that $E \leftrightarrow P$. (A projection such as P is called *cyclic*. The terminology is suggested by that of cyclic groups and the circumstance that the range of P is spanned by the set of all vectors of the form $A^n x$, $n = 1, 2, \cdots$. The reader is advised to supply the proof of this last assertion.) If E is simple, we can exhibit 1 as a cyclic projection by making sure that x has a non-zero component in the range of each E_j; it is true, conversely, that if E is not simple, then 1 is not cyclic.

Although both the characterizations of simplicity described in the two preceding paragraphs have their uses in the infinite-dimensional case, the most revealing and applicable characterization is the one that follows. With each projection P that commutes with E, i.e. with each subspace that reduces A, we associate the least value of the spectral measure E which contains P. This construction has a perfect analog in our column diagram: with each set therein we associate the union of all the columns that have a non-empty intersection with the set. The rows, the objects which enable us to count multiplicities quickly, have an interesting relation to the associated column set. A necessary and sufficient condition that a set be a row (in the sense that it contain not more than one point from each column) is that every one of its subsets may be obtained as the intersection of the given set with a suitable set of columns. Equivalently: a necessary and sufficient condition that the entire diagram consist of but one row is that every one of its subsets be a column. The geometric fact suggested by this characterization is true. A necessary and sufficient condition that a finite-dimensional spectral measure be simple is that every projection P which commutes with it be one of its values. Another way of formulating the same result is this: a necessary and sufficient condition that a finite-dimensional spectral measure be simple is that its values form a maximal abelian set of projections.

If the reader will keep in mind the comments in this section and the preceding one, and if he will systematically compare each definition, each theorem, and each proof with the corresponding concept, assertion, and construction associated with our column diagram, he should have no difficulty in following the remaining technical details. Some of the distinctions that we shall be forced to recognize do not, to be sure, show up in our diagram. If, however, the diagram is generalized so as to admit infinitely many (and possibly even uncountably many) rows and columns, then it becomes an almost perfect schematization of our work. If the countable subsets of the base space, together with their complements, are the ones that are declared measurable, then even the phenomena of non-measurability can be exemplified by generalized column diagrams.

§53. The Commutator of a Set of Projections

From now on we shall again reserve the symbol $\mathfrak{H}$ for an arbitrary but fixed Hilbert space and the word "projection" will refer to projections whose domain is $\mathfrak{H}$.

We recall that the symbol $\leftrightarrow$ denotes commutativity. We now extend its domain of applicability by writing $P \leftrightarrow \mathbf{Q}$ whenever P is an operator, $\mathbf{Q}$ is a set of operators, and $P \leftrightarrow Q$ for all Q in $\mathbf{Q}$. Since we are particularly interested in projections, we introduce the notation $\mathbf{A}$ for the set of all projections, and, if $\mathbf{P}$ is any subset of $\mathbf{A}$, the notation $\mathbf{P}'$ for the set of all those elements P of $\mathbf{A}$ for which $P \leftrightarrow \mathbf{P}$. The purpose of this section is to study the elementary properties of sets such as $\mathbf{P}'$.

THEOREM 1. *If* $\mathbf{P} \subset \mathbf{A}$, *then* $\mathbf{P} \subset \mathbf{P}''$.

THEOREM 2. *If* $\mathbf{P} \subset \mathbf{Q} \subset \mathbf{A}$, *then* $\mathbf{Q}' \subset \mathbf{P}'$.

THEOREM 3. *If* $\mathbf{P} \subset \mathbf{A}$, *then* $\mathbf{P}' = \mathbf{P}'''$.

Proof. Substituting $\mathbf{P}''$ for $\mathbf{Q}$ in Theorem 2, we obtain $\mathbf{P}''' \subset \mathbf{P}'$. If, on the other hand, we apply Theorem 1 to $\mathbf{P}'$ in place of $\mathbf{P}$, we obtain the reverse inequality $\mathbf{P}' \subset \mathbf{P}'''$.

THEOREM 4. *A set* $\mathbf{P}$ *of projections is commutative* (*i.e.* $\mathbf{P} \subset \mathbf{P}'$) *if and only if* $\mathbf{P}''$ *is commutative* (*i.e.* $\mathbf{P}'' \subset \mathbf{P}'$).

Proof. If $\mathbf{P} \subset \mathbf{P}'$, then an application of Theorem 2 shows that $\mathbf{P}'' \subset \mathbf{P}'$. If, on the other hand, $\mathbf{P}'' \subset \mathbf{P}''' = \mathbf{P}'$, then, by Theorem 1, $\mathbf{P} \subset \mathbf{P}'$.

THEOREM 5. *If* $\mathbf{P} \subset \mathbf{A}$, *then* $0 \in \mathbf{P}'$, $1 \in \mathbf{P}'$, *and if* $\{P_j\}$ *is a family of projections in* $\mathbf{P}'$, *then* $\bigvee_j P_j \in \mathbf{P}'$ *and* $\bigwedge_j P_j \in \mathbf{P}'$.

Proof. If P and Q are projections, then $P \leftrightarrow Q$ is equivalent to the assertion that the range of Q is invariant under P. The conclusion follows from the fact that the span and the intersection of any family of subspaces invariant under P are themselves invariant under P.

In view of Theorem 5, the *commutator* $\mathbf{P}'$ of any set $\mathbf{P}$ of projections is a complete sublattice of the lattice $\mathbf{A}$ of all projections. Since $\mathbf{P}'$ contains $1 - P$ along with P, the lattice $\mathbf{P}'$ even possesses a natural complementation operation. It follows that if $\mathbf{P}'$ happens also to be commutative, so that the lattice-theoretic distributive laws are valid in $\mathbf{P}'$ (cf. 30.3), then $\mathbf{P}'$ is a complete Boolean algebra.

§54. Pairs of Commutators

Throughout the remainder of this book $(X, \mathbf{S})$ will be a fixed measurable space and E a fixed spectral measure on $\mathbf{S}$. We shall denote the range of E (i.e. the set of all projections of the form $E(M)$ for some M in $\mathbf{S}$) by $\mathbf{E}$; we shall write $\mathbf{P} = \mathbf{E}'$, and $\mathbf{F} = \mathbf{P}'$. Since $\mathbf{E}$ is commutative, it follows (53.4) that $\mathbf{F} = \mathbf{E}''$ is also commutative and hence that $\mathbf{F}$ is a complete Boolean algebra. The essential relations among $\mathbf{E}$, $\mathbf{F}$, and $\mathbf{P}$ are the inequalities $\mathbf{E} \subset \mathbf{F} \subset \mathbf{P}$ and the equations $\mathbf{E}'' = \mathbf{P}' = \mathbf{F}$ and $\mathbf{F}' = \mathbf{P}$.

The consideration of $\mathbf{F}$ is not one of the things that our heuristic considerations prepared us for; in the finite-dimensional cases $\mathbf{F}$ turns out to be the same as $\mathbf{E}$. In the general case $\mathbf{F}$ may be viewed as a kind of completion of $\mathbf{E}$. The set $\mathbf{E}$ need not be a complete Boolean algebra—$\mathbf{F}$ is. The projections which commute with all the elements of $\mathbf{P} = \mathbf{E}'$ need not belong to $\mathbf{E}$—they do belong to $\mathbf{F}$. Since our development will yield an almost complete insight into the structure of the projections in $\mathbf{F}$, we can only gain information, and not lose any, by incorporating $\mathbf{F}$ into our study.

In all our constructions the space X will play a relatively minor, auxiliary role; what is important is the pair of sets $\mathbf{F}$ and $\mathbf{P}$. We propose, in other words, to present a structure theory for pairs $\mathbf{F}$ and $\mathbf{P}$, where $\mathbf{F}$ and $\mathbf{P}$ are sets of projections, $\mathbf{F}$ is commutative, $\mathbf{F}' = \mathbf{P}$ and $\mathbf{P}' = \mathbf{F}$. Since, however, our proofs will make use of X, E, and $\mathbf{E}$, the material out of which our particular $\mathbf{F}$ and $\mathbf{P}$ were manufactured, it might seem that our promises are greater than our deeds. For the sake of the reader who is interested in the additional generality we record here our assurance that we are not really sacrificing any of it. The point is that the standard theory of representations of Boolean algebras implies that if $\mathbf{F}$ is any complete Boolean algebra of projections (i.e. a complete Boolean subalgebra of **A**), then there exists a measurable space $(X, \mathbf{S})$ and a spectral measure E on **S** such that the range of E is exactly **F**.

In view of the last assertion, the presence in our theory of X, E, and **E** might actually be said to be a gain in generality rather than a loss, since the apparently more general theory involving **F** and **P** alone is always associated with an X and an E such that $\mathbf{E} = \mathbf{F}$. Although the shallowness of this comment is probably obvious, it does help to clarify matters slightly. The various levels of the constructs we will employ are clearer if they are kept separate and if, therefore, it is not assumed that $\mathbf{E} = \mathbf{F}$.

The reader who is not interested in, or did not understand, the preceding two paragraphs, is advised to forget them. Our previews of coming attractions are hereby over, and we are now going to settle down to an uninterrupted showing of the main feature; the cast of characters is, as announced at the beginning of this section, X, E, **E**, **F**, and **P**.

§55. Columns

If $P \in \mathbf{P}$, the *column* generated by P, in symbols $C(P)$, is the smallest element of **F** which contains P: $C(P) = \wedge\{F : P \leqq F \in \mathbf{F}\}$.

The beginning of the theory of columns is quite easy. It is clear, for instance, that $P \leqq C(P)$ for every P in **P** and that $C(P)$ vanishes if and only if P vanishes. It is also clear that the formation of columns is a monotone operation (i.e. that if P and Q are in **P** and $P \leqq Q$, then $C(P) \leqq C(Q)$), and that the column generated by a projection in **F** is itself (i.e. that if $F \in \mathbf{F}$, then $C(F) = F$). On a slightly higher level we encounter the additive and multiplicative properties of the function C.

THEOREM 1. *If $\{P_j\}$ is a family of projections in* **P** *and if*

$$P = \vee_j P_j,$$

then

$$C(P) = \vee_j C(P_j).$$

Proof. Since $P_j \leqq P \leqq C(P)$, it follows that $C(P_j) \leqq C(P)$ for all j and hence $\vee_j C(P_j) \leqq C(P)$. Since, on the other hand, $P_j \leqq C(P_j)$ for all j, we have also $P = \vee_j P_j \leqq \vee_j C(P_j)$ and consequently

$$C(P) \leqq \vee_j C(P_j).$$

(Recall that $\vee_j C(P_j) \in \mathbf{F}$.)

There is no intuitive geometric reason for expecting C to be multiplicative as well as additive, and indeed it is not; the following result exhibits the one little shred of multiplicative behaviour that C does possess.

THEOREM 2. *If* $P \in \mathbf{P}$ *and* $F \in \mathbf{F}$, *then* $C(FP) = FC(P)$.

Proof. Since $FP \leqq F$ and $FP \leqq P$, it follows that

$$C(FP) \leqq C(F) = F$$

and

$$C(FP) \leqq C(P),$$

and consequently that $C(FP) \leqq FC(P)$. Since, on the other hand,

$$P = (1 - F)P + FP \leqq (1 - F) + FP,$$

it follows that

$$C(P) \leqq (1 - F) + C(FP),$$

and hence that

$$FC(P) \leqq FC(FP) \leqq C(FP).$$

Because of its later applicability we record here for reference an immediate corollary of Theorem 2.

THEOREM 3. *If* $P \in \mathbf{P}$, $F \in \mathbf{F}$, *and* $0 \neq F \leqq C(P)$, *then* $FP \neq 0$.

Proof. $C(FP) = FC(P) = F$.

§56. Rows

A *row* is a projection R in $\mathbf{P}$ such that if $R \geqq P \in \mathbf{P}$, then $P = C(P)R$. We note that if P and R are projections in $\mathbf{P}$ such that $P \leqq R$, then, since $P \leqq C(P)$, the inequality $P \leqq C(P)R$ is always valid. The statement that R is a row means that the inequality reduces to an equality for all admissible P.

THEOREM 1. *If* R *is a row and if* $R \geqq S \in \mathbf{P}$, *then* S *is a row.*

Proof. If $S \geqq P \in \mathbf{P}$, then $R \geqq P$ and therefore $P = C(P)R$; it follows that $P = PS = C(P)R \cdot S = C(P)S$.

THEOREM 2. *If* R *is a row and if* P *and* Q *are projections in* $\mathbf{P}$ *such that* $P \leqq R$ *and* $Q \leqq R$, *then* $P \leftrightarrow Q$ *and* $C(PQ) = C(P)C(Q)$. *If*

$$C(P) \leqq C(Q),$$

then

$$P \leqq Q.$$

Proof. Since $P = C(P)R$ and $Q = C(Q)R$, the commutativity of P and Q and the last assertion of the theorem are obvious. To prove that under these special circumstances C is multiplicative, we note that since $C(P) \leqq C(R)$ and $C(Q) \leqq C(R)$, it follows that

$$C(P)C(Q)C(R) = C(P)C(Q).$$

The desired conclusion follows from an application of 55.2 to the relation $PQ = C(P)C(Q)R$.

Everything we are going to do from here on in will aim at showing how rows are always put together to form columns. At the present stage, however, our discussion is somewhat hampered by the fact that we have no particular reason to believe that such things as rows even exist. We find it necessary, therefore, to begin a somewhat lengthy detour whose purpose is to dig out the rows that we need from the Hilbert space and the spectral measure that are at the basis of our theory.

§57. Cycles

For any vector x in $\mathfrak{H}$, the *cyclic projection* or more concisely the *cycle* generated by x, in symbols $Z(x)$, is the projection on the subspace of $\mathfrak{H}$ spanned by the set of all vectors of the form $E(M)x$, $M \in \mathbf{S}$. Our first duty is to show that the concept of cycle is not entirely foreign to the subject we are studying.

THEOREM 1. *If $x \in \mathfrak{H}$, then $Z(x) \in \mathbf{P}$.*

Proof. If M and N are in $\mathbf{S}$, then $E(M)E(N)x = E(M \cap N)x$, so that the range of $Z(x)$ is invariant under $E(M)$. It follows that the range of $Z(x)$ reduces $E(M)$, and hence that $E(M) \leftrightarrow Z(x)$. Since M is arbitrary, this means that $Z(x) \in \mathbf{E}' = \mathbf{P}$.

We can now proceed with good conscience to derive the properties of cycles and their relations to rows and columns.

THEOREM 2. *If $P \in \mathbf{P}$, $x \in \mathfrak{H}$, and $P \leqq Z(x)$, then $P = Z(Px)$.*

Proof. The range of $Z(Px)$ is, by definition, the span of the set of all vectors of the form $E(M)Px = PE(M)x$, $M \in \mathbf{S}$. It follows that the range of $Z(Px)$ is the image under P of the span of the set of all vectors of the form $E(M)x$, $M \in \mathbf{S}$, and hence that $Z(Px) = PZ(x) = P$.

THEOREM 3. *If $F \in \mathbf{F}$ and $x \in \mathfrak{H}$, then $FZ(x) = Z(Fx)$.*

Proof. If $P = FZ(x)$, then, by Theorem 2,

$$FZ(x) = P = Z(Px) = Z(FZ(x)x) = Z(Fx).$$

THEOREM 4. *If* $P \in \mathbf{P}$ *and* $x \in \mathfrak{H}$, *then* $Px = 0$ *and* $Px = x$ *are equivalent to* $Z(x) \perp P$ *and* $Z(x) \leqq P$ *respectively.*

Proof. If $M \in \mathbf{S}$ and $Px = 0$, then $PE(M)x = E(M)Px = 0$. It follows that $Py = 0$ whenever y belongs to the range of $Z(x)$ and hence that $PZ(x) = 0$. Applying this result to $1 - P$ in place of P, we see that $Z(x) \leqq P$ whenever x belongs to the range of P. These remarks prove a half of both the asserted equivalences; the remaining halves are trivial.

It is time to observe that if $x = 0$, then $Z(x) = 0$, and that the more significant converse of this implication is also valid. (Recall that the range of $Z(x)$ contains $E(X)x = x$.) If we introduce the convenient abbreviation $C(x)$ for $C(Z(x))$, then we can announce a similar state-about $C(x)$: a necessary and sufficient condition that $C(x) = 0$ is that $x = 0$. This last assertion has a slight generalization which we shall find useful.

THEOREM 5. *If* $F \in \mathbf{F}$, $x \in \mathfrak{H}$, *and* $0 \neq F \leqq C(x)$, *then* $Fx \neq 0$.

Proof. If $Fx = 0$, then, by Theorem 4, $FZ(x) = 0$ and therefore $F = F \cdot C(x) = C(F \cdot Z(x)) = 0$.

§58. Separable Projections

A projection F in $\mathbf{F}$ is *separable* if every orthogonal family $\{F_j\}$ of non-zero projections in $\mathbf{F}$, such that $F_j \leqq F$ for all j, is necessarily countable. The main purpose of this section is to show that the columns $C(x)$, introduced at the end of the preceding section, are intrinsically characterized by the property of separability. We observe that if F and G are projections in $\mathbf{F}$ such that $F \leqq G$ and G is separable, then F is separable.

THEOREM 1. *If* $\{F_j\}$ *is a countable orthogonal family of separable projections in* $\mathbf{F}$, *and if* $F = \bigvee_j F_j$, *then* F *is separable.*

Proof. If $\{G_k\}$ is an orthogonal family of non-zero projections in $\mathbf{F}$ such that $G_k \leqq F$ for all k, then $\{F_j G_k\}$ is, for each value of j, an orthogonal family of projections in $\mathbf{F}$ such that $F_j G_k \leqq F_j$ for all k. It follows that, for each j, $F_j G_k = 0$ except for a countable set of values of k. Since $G_k = FG_k = \bigvee_j F_j G_k$ for all k, it follows that $\{G_k\}$ is countable.

THEOREM 2. *If* $x \in \mathfrak{H}$, *then* $C(x)$ *is separable.*

Proof. If $\{F_j\}$ is an orthogonal family of non-zero projections in **F** such that $F_j \leqq C(x)$ for all j, then $\Sigma_j F_j \leqq C(x)$ and consequently $\Sigma_j \| F_j x \|^2 \leqq \| x \|^2$. The countability of $\{F_j\}$ follows from 57.5.

THEOREM 3. *If P and Q are in* **P**, *if $P \leqq C(Q)$, and if $C(Q)$ is separable, then there exists a vector x in the range of P such that $C(P) = C(x)$. Hence, in particular, if $F \epsilon$ **F** and F is separable, then $F = C(x)$ for some vector x.*

Proof. Let $\{x_j\}$ be a maximal family of non-zero vectors in the range of P such that $C(x_j)C(x_k) = 0$ whenever $j \neq k$. Since $C(x_j) \leqq C(Q)$ for all j, and since $C(Q)$ is separable, the family $\{x_j\}$ is countable and there is therefore no loss of generality in assuming that $\Sigma_j \| x_j \|^2 < \infty$. If we write $x = \Sigma_j x_j$, then x is in the range of P; we shall complete the proof by showing that $C(P) = C(x)$. If $C(P) - C(x) \neq 0$, then, by 55.3, the range of $(C(P) - C(x))P$ contains a non-zero vector y. It follows that y belongs to the range of P and hence to the range of $C(P)$. Since y also belongs to the range of $C(P) - C(x)$, it follows that $C(x)y = 0$. Using 57.4, we see that $C(x)Z(y) = 0$ and hence, by 55.2, that $C(x)C(y) = 0$. If we knew that $C(x_j) \leqq C(x)$ for all j, then we could conclude that the existence of y contradicts the maximality of the family $\{x_j\}$, and the proof would be complete. It is therefore sufficient to prove that $Z(x_j) \leqq Z(x)$ for all j.

Since $C(x_j)x_k = \delta_{jk}x_k$, it follows that $C(x_j)x = x_j$ and hence (since $C(x_j) \epsilon \mathbf{F}$) that $Z(x)x_j = Z(x)C(x_j)x = C(x_j)Z(x)x = C(x_j)x = x_j$ for all j. Consequently $Z(x)E(M)x_j = E(M)Z(x)x_j = E(M)x_j$ whenever $M \epsilon \mathbf{S}$, and therefore $Z(x)Z(x_j) = Z(x_j)$ or $Z(x_j) \leqq Z(x)$ for all j.

§59. Characterizations of Rows

Since on several occasions we shall run into pairs of projections P and Q in **P** such that $C(P)C(Q) = 0$, it is convenient to introduce a technical term for the phenomenon; under these circumstances we shall say that P and Q are *very orthogonal.*

THEOREM 1. *A necessary and sufficient condition that a projection R in* **P** *be a row is that if $R \geqq P \epsilon$* **P**, *$R \geqq Q \epsilon$* **P**, *and P and Q are orthogonal, then P and Q are very orthogonal.*

Proof. If R is a row, if $P = C(P)R$, $Q = C(Q)R$, and if $PQ = 0$, then $C(P)C(Q)R = 0$. Applying 55.2, we conclude that

$$C(P)C(Q)C(R) = 0,$$

and hence, by the monotony of the formation of columns, $C(P)C(Q) = 0$. This proves the necessity of the condition; to prove sufficiency we suppose that $R \geqq P \epsilon \mathbf{P}$ and we write $Q = C(P)R - P$. It is clear that $R \geqq Q \epsilon \mathbf{P}$ and that $PQ = 0$; it follows from the hypothesis that $C(P)C(Q) = 0$. Since, however, the relation $Q \leqq C(P)$ implies that $C(Q) \leqq C(P)$, we may conclude that $C(Q) = 0$ and hence that $Q = 0$. In other words $P = C(P)R$, and therefore, since P is arbitrary, R is a row.

We turn next to one of the results whose object is to tie together the various concepts we have introduced. We shall be able to make use of the result immediately to obtain (in Theorem 3 below) a significant strengthening of Theorem 1.

THEOREM 2. *If $P \epsilon \mathbf{P}$, then there exists an orthogonal family $\{Z(x_j)\}$ of cycles such that $P = \bigvee_j Z(x_j)$.*

Proof. Let $\{x_j\}$ be a maximal family of non-zero vectors in the range of P such that $Z(x_j)Z(x_k) = 0$ whenever $j \neq k$. If

$$P - \bigvee_j Z(x_j) \neq 0,$$

then the range of P contains a non-zero vector x such that $Z(x_j)x = 0$ for all j. It follows from 57.4 that $Z(x_j)Z(x) = 0$ for all j. Since this contradicts the assumed maximality of the family $\{x_j\}$, we must have $P - \bigvee_j Z(x_j) = 0$.

In view of our subsequent results on orthogonal sums of cycles, the reader is warned to make an effort to keep straight the conclusion of Theorem 2. The essential point is that the family $\{Z(x_j)\}$ is not asserted to be very orthogonal.

THEOREM 3. *A necessary and sufficient condition that a projection R in $\mathbf{P}$ be a row is that if $Z(x)$ and $Z(y)$ are orthogonal cycles such that $R \geqq Z(x)$ and $R \geqq Z(y)$, then $Z(x)$ and $Z(y)$ are very orthogonal.*

Proof. The necessity of the condition follows from Theorem 1. To prove sufficiency, we suppose that $R \geqq P \epsilon \mathbf{P}$, $R \geqq Q \epsilon \mathbf{P}$, and P and Q are orthogonal; in view of Theorem 1, the desideratum is to prove that P and Q are very orthogonal. According to Theorem 2, there exist orthogonal families $\{Z(x_j)\}$ and $\{Z(y_k)\}$ of cycles such that $P = \bigvee_j Z(x_j)$ and $Q = \bigvee_k Z(y_k)$. Since $PQ = 0$, it follows that

$$Z(x_j)Z(y_k) = 0$$

for all j and k and therefore, by the hypothesis of the theorem,

$$C(x_j)C(y_k) = 0$$

for all j and k. Since the additivity of the function C implies that

$$C(P) = \vee_j C(x_j)$$

and

$$C(Q) = \vee_k C(y_k),$$

we may conclude that $C(P)C(Q) = 0$.

§60. Cycles and Rows

Our theorems get deeper all the time. In this section we prove two key propositions, the first of which asserts that cycles do indeed have the measure-theoretic characterization that our heuristic comments hinted at. (The second one can speak for itself.)

THEOREM 1. *If $x \in \mathfrak{H}$ and if μ is the measure on* **S** *defined for every M in* **S** *by $\mu(M) = (E(M)x, x)$, then there exists an isomorphism U from $\mathfrak{L}_2(\mu)$ onto the range of $Z(x)$ such that $U^{-1}E(M)Uf = \chi_M \cdot f$ whenever $f \in \mathfrak{L}_2(\mu)$ and $M \in$* **S**.

Proof. We write $U\chi_M = E(M)x$ for every M in **S**. If the definition of U is extended from characteristic functions to simple functions by the requirement of linearity, then, in view of the definition of $Z(x)$, U becomes a linear transformation from a dense subset of $\mathfrak{L}_2(\mu)$ onto a dense subset of the range of $Z(x)$. The additivity of E guarantees the uniqueness of the definition of U. Since the relations $\| \chi_M \|^2 = \mu(M) = (E(M)x, x) = \| E(M)x \|^2 = \| U\chi_M \|^2$ shows that U is norm-preserving, U may be extended to an isomorphism. If M_0 and M are in **S**, then

$$U(\chi_{M_0} \cdot \chi_M) = U(\chi_{M_0 \cap M}) = E(M_0 \cap M)x$$
$$= E(M_0)E(M)x = E(M_0)U\chi_M .$$

This means that $U(\chi_{M_0} \cdot f) = E(M_0)Uf$ whenever $f = \chi_M$; approximation by simple functions proves the validity of the relation for all f in $\mathfrak{L}_2(\mu)$.

THEOREM 2. *Every cycle is a row.*

Proof. We are to prove that if $x \in \mathfrak{H}$ and if $Z(x) \geqq P \in \mathbf{P}$, then $P = C(P)Z(x)$. It is convenient to use the result and the notation of Theorem 1. If $Q = U^{-1}PU$, then Q is a projection with domain $\mathfrak{L}_2(\mu)$. If $M \in$ **S**, then, for every f in $\mathfrak{L}_2(\mu)$,

$$Q(\chi_M \cdot f) = U^{-1}PU(\chi_M \cdot f) = U^{-1}PE(M)Uf = U^{-1}E(M)PUf$$
$$= U^{-1}E(M)U \cdot U^{-1}PUf = \chi_M \cdot Qf.$$

Consider in particular the function h identically equal to 1 and write $\chi = Qh$. Since $\chi \cdot \chi_M = \chi_M \cdot \chi = \chi_M \cdot Qh = Q(\chi_M \cdot h) = Q\chi_M$, i.e. since Q and multiplication by χ have the same effect on every χ_M, it follows that $\chi \cdot f = Qf$ for all f in $\mathfrak{L}_2(\mu)$. The fact that Q is idempotent implies that our notation is justified, i.e. that $\chi = \chi_{M_0}$ for some M_0 in $\mathbf{S}$.

If now y is any vector in the range of $Z(x)$, so that $y = Uf$ for some f in $\mathfrak{L}_2(\mu)$, then $Py = PUf = UQf = U(\chi_{M_0} \cdot f) = E(M_0)Uf = E(M_0)y$. The arbitrariness of y implies that $P = PZ(x) = E(M_0)Z(x)$. It follows that

$$C(P) = E(M_0)C(x) \leqq E(M_0).$$

Since

$$P \leqq C(P)Z(x) \leqq E(M_0)Z(x) = P,$$

the proof is complete.

§61. The Existence of Rows

Our detour is almost over. The last result that we obtained shows that rows exist and even (in view of 59.2) that they exist in abundance. The main purpose of this section is to prove, on the basis of a couple of preliminary results, that there exist rows of arbitrarily prescribed lengths.

THEOREM 1. *If* $\{Z_j\}$ *is a very orthogonal family of cycles and if* $R = \bigvee_j Z_j$, *then* R *is a row.*

Proof. Suppose that $Z(x)$ and $Z(y)$ are orthogonal cycles such that $R \geqq Z(x)$ and $R \geqq Z(y)$; in virtue of 59.3 it is sufficient to prove that $Z(x)$ and $Z(y)$ are very orthogonal. If we write

$$x_j = C(y)C(Z_j)x,$$

then 57.3 implies that $Z(x_j) = Z(x)C(y)C(Z_j)$ for all j; if, similarly,

$$y_j = C(x)C(Z_j)y,$$

then $Z(y_j) = C(x)Z(y)C(Z_j)$ for all j. Since

$$Z(x_j) \leqq Z(x)$$

and

$$Z(y_j) \leqq Z(y)$$

for all j, it follows from the orthogonality we have assumed, that

$$Z(x_j)Z(y_j) = 0$$

for all j. On the other hand we have

$$C(x_j) = C(x)C(y)C(Z_j) = C(y_j)$$

for all j. Since x_j belongs to the range of $C(Z_j)$, it is orthogonal to the range of Z_k whenever $k \neq j$ and, consequently, x_j belongs to the range of Z_j; similarly, of course, y_j belongs to the range of Z_j. Since Z_j is a row (cf. 60.2), it follows from 56.2 that $Z(x_j) = Z(y_j)$ for all j. The only way to reconcile our apparently contradictory results is to conclude that all $Z(x_j)$ and $Z(y_j)$ and therefore all $C(x)C(y)C(Z_j)$ vanish. Since $C(x)C(y) = C(x)C(y)C(R) = \vee_j C(x)C(y)C(Z_j) = 0$, we have proved what we had to prove.

THEOREM 2. *If $\{R_j\}$ is a very orthogonal family of rows and if*

$$R = \vee_j R_j,$$

then R is a row.

Proof. Using 59.2, we may express each R_j as an orthogonal sum of cycles. The fact that each R_j is a row implies that any two distinct ones of its summands are very orthogonal. The fact that $\{R_j\}$ is a very orthogonal family implies that if $j \neq k$, then any summand of R_j is very orthogonal to any summand of R_k. If, in other words, we unite into one family all the cycles used to obtain all R_j, we obtain a representation of R as a very orthogonal sum of cycles, and Theorem 2 becomes an immediate corollary of Theorem 1.

THEOREM 3. *If $P \in \mathbf{P}$, then there exists a row R such that $R \leqq P$ and $C(R) = C(P)$.*

Proof. Let $\{R_j\}$ be a maximal very orthogonal family of non-zero rows such that $R_j \leqq P$ for all j. If it is not true that $P \leqq \vee_j C(R_j)$, then (since $P \leftrightarrow \vee_j C(R_j)$) there exists a non-zero vector x in the range of P such that $C(R_j)x = 0$ for all j. Since $Z(x)$ is a row, $Z(x) \leqq P$, and since, by 57.4 and 55.2, $C(R_j)C(x) = 0$ for all j, the existence of x contradicts the maximality of the family $\{R_j\}$. We are therefore forced to accept the inequality $P \leqq \vee_j C(R_j)$ and, as a consequence, the inequality $C(P) \leqq \vee_j C(R_j) = C(\vee_j R_j)$. Since the reverse of the last-written inequality is obvious and since, by Theorem 2, $\vee_j R_j$ is a row, the proof is complete.

§62. Orthogonal Systems

If $F \in \mathbf{F}$, an *orthogonal system* of type F is an orthogonal family $\{R_j\}$ of non-zero rows such that $C(R_j) = F$ for all j. The purpose of this section is to show how to construct various orthogonal systems and how to put together the ones we have constructed to obtain bigger ones.

THEOREM 1. *If u is a cardinal number, if $\{F_j\}$ is an orthogonal family of non-zero projections in* **F**, *and if, for each j, $\{R_{jk}\}$ is an orthogonal system of type F_j and of power u, then $\{\vee_j R_{jk}\}$ is an orthogonal system of type $\vee_j F_j$ and of power u.*

Proof. If, for each index k, $R_k = \vee_j R_{jk}$, then it follows from 61.2 that $\{R_k\}$ is an orthogonal family of non-zero rows. The proof is completed by the observation that $C(R_k) = \vee_j C(R_{jk}) = \vee_j F_j$ for all k.

We observe that if $\{R_j\}$ is an orthogonal system of type F, then $\vee_j R_j \leqq F$; the orthogonal system $\{R_j\}$ is called *complete* if $\vee_j R_j = F$. It is obvious that a complete orthogonal system of type F is a maximal orthogonal system of type F; we shall presently see that every maximal orthogonal system of type F is put together from complete orthogonal systems of suitable types.

THEOREM 2. *If $\{R_j\}$ is an orthogonal system of type F and if F_0 is a non-zero projection in* **F** *such that $F_0 \leqq F$, then $\{F_0 R_j\}$ is an orthogonal system of type F_0; if $\{R_j\}$ is complete, then so is $\{F_0 R_j\}$.*

Proof. It is clear that $\{F_0 R_j\}$ is an orthogonal family of projections in **P** and that $C(F_0 R_j) = F_0 F = F_0$ for all j. Since, for all j, $F_0 R_j \neq 0$ (by 55.3) and $F_0 R_j$ is a row (by 56.1), it follows that $\{F_0 R_j\}$ is indeed an orthogonal system of type F_0. If $\vee_j R_j = F$, then

$$\vee_j F_0 R_j = F_0 \cdot \vee_j R_j = F_0 .$$

THEOREM 3. *If $\{R_j\}$ is an orthogonal system of type F and if F_0 is a non-zero projection in* **F** *such that $F_0 \leqq \vee_j R_j$, then $\{F_0 R_j\}$ is a complete orthogonal system of type F_0.*

Proof. In view of Theorem 2 it is sufficient to prove completeness, and this is a consequence of the relations $F_0 = F_0 \cdot \vee_j R_j = \vee_j F_0 R_j$.

THEOREM 4. *If $\{R_j\}$ is a maximal orthogonal system of (necessarily non-zero) type F, then there exists a vector x in the range of F such that $\{C(x) R_j\}$ is a complete orthogonal system of type $C(x)$.*

Proof. If $P = F - \vee_j R_j$, then, since $P \leqq F$, it follows that

$$C(P) \leqq F.$$

If $C(P) = F$, then, by 61.3, there exists a row R such that $R \leqq P$ and $C(R) = F$. Since this contradicts the maximality of $\{R_j\}$, it follows that $F_0 = F - C(P) \neq 0$. Since the relation $F_0 C(P) = 0$ implies that

$$F_0 P = 0,$$

and since this in turn implies (in view of the definition of P) that

$$F_0 \leqq \mathsf{V}_j R_j,$$

it follows from Theorem 3 that $\{F_0 R_j\}$ is a complete orthogonal system of type F_0. The proof of Theorem 4 may be completed by selecting an arbitrary non-zero vector in the range of F_0 and applying Theorem 2 to $\{F_0 R_j\}$, F_0, and $C(x)$ in place of $\{R_j\}$, F, and F_0 respectively.

§63. The Power of a Maximal Orthogonal System

The theorem of the present section is the fundamental theorem of multiplicity theory.

THEOREM 1. *If $F \in \mathbf{F}$, then any two maximal orthogonal systems of type F have the same power.*

Proof. If $F = 0$, there are no orthogonal systems of type F and the power in question is zero. Suppose then that u and v are non-zero cardinal numbers and that $\{R_j\}$ and $\{S_k\}$ are maximal orthogonal systems of type F and of power u and v respectively. By symmetry it will be sufficient to prove that $v \leqq u$.

By 62.4, there exists a non-zero vector x in the range of F such that $\{C(x)R_j\}$ is a complete orthogonal system of type $C(x)$. Since we may replace F, $\{R_j\}$, and $\{S_k\}$ by $C(x)$, $\{C(x)R_j\}$, and $\{C(x)S_k\}$ respectively (cf. 62.2), we may (and do) assume that $\{R_j\}$ and $\{S_k\}$ are orthogonal systems of type $C(x)$ and of power u and v respectively, and that

$$\mathsf{V}_k S_k \leqq \mathsf{V}_j R_j;$$

under these conditions we shall prove that $v \leqq u$.

Since $R_j \leqq C(R_j) = C(x)$ for all j, it follows from 58.2 and 58.3 that $R_j = Z(x_j)$ for a suitable vector x_j; similarly we may find, for each k, a vector y_k such that $S_k = Z(y_k)$.

Suppose now that u is infinite. For each value of j, let K_j be the set of those indices k for which $Z(y_k)x_j \neq 0$; it is clear that each K_j is countable. If $k \in' \mathsf{U}_j K_j$, i.e. if $Z(y_k)x_j = 0$ for all j, then $Z(y_k)Z(x_j) = 0$ for all j and therefore $Z(y_k) = Z(y_k) \cdot \mathsf{V}_j Z(x_j) = 0$. Since this is false, it follows that every k belongs to $\mathsf{U}_j K_j$ and hence that $v \leqq \aleph_0 \cdot u = u$.

In case u is finite, the proof is a bit more complicated. For each index j we write μ_j for the measure on $\mathbf{S}$ defined for every M in $\mathbf{S}$ by

$$\mu_j(M) = (E(M)x_j, x_j).$$

According to 60.1, there exists an isomorphism U_j from $\mathfrak{L}_2(\mu_j)$ onto the range of $Z(x_j)$ such that $U_j^{-1}E(M)U_j f_j = \chi_M \cdot f_j$ whenever $f_j \in \mathfrak{L}_2(\mu_j)$ and

$M \in \mathbf{S}$. Putting together the separate isomorphisms U_j we obtain an isomorphism U from the direct sum, say $\mathfrak{K}$, of all $\mathfrak{L}_2(\mu_j)$ onto the range of $\bigvee_j Z(x_j)$ such that $U^{-1}E(M)U\{f_j\} = \{\chi_M \cdot f_j\}$ whenever $\{f_j\} \in \mathfrak{K}$ and $M \in \mathbf{S}$.

We define a measure μ on $\mathbf{S}$ by writing $\mu(M) = (E(M)x, x)$ for every M in $\mathbf{S}$. If $\mu(M) = 0$ for some M, then $E(M)x = 0$ and therefore $E(M)C(x) = 0$. Since $C(x) = C(x_j)$, it follows that $E(M)C(x_j) = 0$ and therefore, in particular, $E(M)x_j = 0$ for every j. These considerations imply that each of the measures μ_j is absolutely continuous with respect to μ and that, therefore, there exists a family $\{g_j\}$ of non-negative functions in $\mathfrak{L}_1(\mu)$ such that $\mu_j(M) = \int_M g_j \, d\mu$ for all j and for every M in $\mathbf{S}$.

Since y_k belongs to the range of $\bigvee_j Z(x_j)$ for all k, we can find vectors $\{f_{jk}\}$ in $\mathfrak{K}$ such that $y_k = U\{f_{jk}\}$. If $M \in \mathbf{S}$, then

$$\begin{aligned}(E(M)y_{k_1}, y_{k_2}) &= (E(M)U\{f_{jk_1}\}, U\{f_{jk_2}\}) = (U\{\chi_M \cdot f_{jk_1}\}, U\{f_{jk_2}\}) \\ &= (\{\chi_M \cdot f_{jk_1}\}, \{f_{jk_2}\}) = \Sigma_j \int \chi_M f_{jk_1} f^*_{jk_2} \, d\mu_j = \Sigma_j \int_M f_{jk_1} f^*_{jk_2} g_j \, d\mu \\ &= \int_M \Sigma_j f_{jk_1} f^*_{jk_2} g_j \, d\mu.\end{aligned}$$

If $\mu(M) \neq 0$, then a repetition of the argument of the preceding paragraph shows that a necessary and sufficient condition for the vanishing of $(E(M)y_{k_1}, y_{k_2})$ is that $k_1 \neq k_2$. It follows that if k_1 and k_2 are restricted to a countable subset of the index set $\{k\}$, then there exists a set M in $\mathbf{S}$ such that $\mu(M) = 0$ and such that if $t \in X - M$, then a necessary and sufficient condition for the vanishing of

$$\Sigma_j f_{jk_1}(t) \, f^*_{jk_2}(t) \, g_j(t)$$

is that $k_1 \neq k_2$. Since for a fixed t in $X - M$, and for each k, $\{f_{jk}(t)\}$ is a vector in a u-dimensional Hilbert space in which, therefore, the power of an orthogonal set of non-zero vectors is not greater than u, it follows that indeed $v \leqq u$, and the proof is complete.

§64. Multiplicities

The result of the preceding section enables us to associate a unique cardinal number with every projection F in $\mathbf{F}$. We define the *multiplicity* of F, in symbols $u(F)$, to be the power (possibly zero) of a maximal orthogonal system of type F. The function u, from $\mathbf{F}$ to cardinal numbers, behaves very much like the multiplicity functions we defined in §49.

THEOREM 1. *If F and G are projections in* $\mathbf{F}$ *such that* $0 \neq F \leqq G$, *then* $u(G) \leqq u(F)$; *if* $F = 0$, *then* $u(F) = 0$.

Proof. If $\{R_j\}$ is an orthogonal system of type G, then, by 62.2, $\{FR_j\}$ is an orthogonal system of type F. This proves the first assertion; the second assertion is obvious.

THEOREM 2. *If* $\{F_j\}$ *is an orthogonal family of non-zero projections in* $\mathbf{F}$ *and if* $F = \bigvee_j F_j$, *then* $u(F) = \min\{u(F_j)\}$.

Proof. We write $u = \min\{u(F_j)\}$. Since $F_j \leqq F$ for all j, it follows from Theorem 1 that $u(F) \leqq u(F_j)$ for all j and hence that $u(F) \leqq u$. Since, on the other hand, $u(F_j) \geqq u$ for all j, it follows that, for each j, there exists an orthogonal system $\{R_{jk}\}$ of type F_j and of power u. Since, by 62.1, $\{\bigvee_j R_{jk}\}$ is an orthogonal system of type F, it follows that $u(F) \geqq u$.

We continue imitating the theory of multiplicity functions. If a projection F in $\mathbf{F}$ is such that $u(F) = u(F_0)$ whenever F_0 is a non-zero projection in $\mathbf{F}$ such that $F_0 \leqq F$, we shall say that F has *uniform multiplicity*.

THEOREM 3. *If* $\{F_j\}$ *is an orthogonal family of projections in* $\mathbf{F}$ *such that each* F_j *has uniform multiplicity* u, *and if* $F = \bigvee_j F_j$, *then* F *has uniform multiplicity* u.

Proof. If F_0 is a non-zero projection in $\mathbf{F}$ such that $F_0 \leqq F$, then $F_0 = \bigvee_j F_0 F_j$. Since the last-written equation remains valid if the supremum is extended over those indices j for which $F_0 F_j \neq 0$, it follows from the uniformity of the F_j's and from Theorem 2, that $u(F_0) = u$.

THEOREM 4. *A necessary and sufficient condition that a non-zero projection* F *in* $\mathbf{F}$ *have uniform multiplicity is that there exist a complete orthogonal system of type* F.

Proof. The sufficiency of the condition follows, using 62.2, from the fact that a complete orthogonal system is maximal. To prove its necessity, we let $\{F_j\}$ be a maximal orthogonal family of non-zero projections in $\mathbf{F}$ such that $F_j \leqq F$ for all j and such that for each j there exists a complete orthogonal system of type F_j. That such families exist, and that, in fact, the maximality of $\{F_j\}$ implies $\bigvee_j F_j = F$, follows from 62.4. Since the power of a complete orthogonal system of type F_j is exactly $u(F)$ for all j, it is legitimate to denote such a system by $\{R_{jk}\}$, with the same index set $\{k\}$ for all j. If $R_k = \bigvee_j R_{jk}$, then, by 62.1, $\{R_k\}$ is an orthogonal system of type F; the completeness of $\{R_k\}$ follows from the relations $\bigvee_k R_k = \bigvee_j \bigvee_k R_{jk} = \bigvee_j F_j = F$.

THEOREM 5. *If, for each cardinal number u not exceeding the dimension of $\mathfrak{H}$, F_u is the supremum of all those projections in* **F** *which have uniform multiplicity u, then $\{F_u\}$ is an orthogonal family, $\vee_u F_u = 1$, and, for each u, either $F_u = 0$ or F_u has uniform multiplicity u.*

Proof. For a fixed cardinal number u, let $\{G_j\}$ be a maximal orthogonal family of projections in **F** such that each G_j has uniform multiplicity u. If $G = F_u - \vee_j G_j \neq 0$, then there exists a projection F in **F** such that F has uniform multiplicity u and such that $FG \neq 0$. Since FG has uniform multiplicity u, this contradicts the maximality of the family $\{G_j\}$. Consequently $F_u = \vee_j G_j$, and therefore either $F_u = 0$ or, by Theorem 3, F_u has uniform multiplicity u. It follows that if $F_u F_v \neq 0$, then, since $F_u F_v \leqq F_u$ and $F_u F_v \leqq F_v$, the multiplicity of $F_u F_v$ is equal to u and to v at the same time, or, in other words, $u = v$. The fact that $1 - \vee_u F_u = 0$ follows from 62.4 and Theorem 4.

The results of this section essentially conclude the structure theory of the pair of sets **F** and **P**. Theorem 5 shows us that $\mathfrak{H}$ decomposes in a natural and intrinsically defined manner into pieces of uniform multiplicity; Theorem 4 tells us that each such piece is made up of rows cutting all the way across. From 59.2 we know that every projection in **P** and therefore, in particular, each of the rows that make up one of the uniform pieces, is an orthogonal sum of cycles; according to 60.1, the given spectral measure behaves on each such cycle as do the multiplications by characteristic functions of measurable sets on a finite measure space. In the remaining sections we tie this all up with multiplicity functions so as to obtain the isomorphism of E with a canonical spectral measure.

§65. Measures from Vectors

If x is a vector in $\mathfrak{H}$, we shall write $\rho(x)$ for the measure μ defined for every M in **S** by $\mu(M) = (E(M)x, x)$. In this section we shall study the relation of the function ρ to some of the other concepts we have introduced. The first and most obvious property of ρ is that $\rho(x) = 0$ if and only if $x = 0$; for the proof we need merely to recall that since $E(X) = 1$, it follows that $(E(X)x, x) = \| x \|^2$. A slightly less obvious property of ρ is a kind of additivity: if $\{Z(x_j)\}$ is an orthogonal family of cycles, and if the family $\{x_j\}$ of vectors is summable with sum x, then

$$\rho(x) \equiv \vee_j \rho(x_j).$$

To prove this we observe that, for each value of j, $E(M)x_j$ belongs to the range of $Z(x_j)$ for every M in **S**; it follows that $\{E(M)x_j\}$ is an orthogonal family of vectors and that $\| E(M)x \|^2 = \Sigma_j \| E(M)x_j \|^2$ for every M in **S**. Our next result lies somewhat deeper.

THEOREM 1. *If x and y are vectors, a necessary and sufficient condition for the orthogonality of $\rho(x)$ and $\rho(y)$ is the orthogonality of $C(x)$ and $C(y)$.*

Proof. We write $\mu = \rho(x)$ and $\nu = \rho(y)$. If $\mu \perp \nu$, then there exists a set M in **S** such that $\nu(M) = \mu(X - M) = 0$. (There are many ways of seeing this: one way is to apply 47.2 first to ν and $\nu \vee \mu$ and then to μ and $\nu \vee \mu$.) It follows that $E(M)y = E(X - M)x = 0$ and hence that $y = E(X - M)y$ and $x = E(M)x$. Since this implies that

$$C(y) \leqq E(X - M)$$

and

$$C(x) \leqq E(M),$$

the orthogonality of $C(x)$ and $C(y)$ follows from that of $E(M)$ and $E(X - M)$.

Suppose now that we know that $C(x)$ and $C(y)$ are orthogonal. Since $C(Z(x) \vee Z(y)) = C(x) \vee C(y)$ and since, by 58.1 and 58.2, $C(x) \vee C(y)$ is separable, it follows from 58.3 that there exists a vector z in the range of $Z(x) \vee Z(y)$ such that $C(x) \vee C(y) = C(z)$. Write $\mu = \rho(z)$ and let U be the isomorphism described in 60.1 from $\mathfrak{L}_2(\mu)$ onto the range of $Z(z)$. If $x = Uf$ and $y = Ug$, then, since $Z(x)Z(y) = 0$, it follows that

$$0 = (E(M)x, y) = (E(M)Uf, Ug) = (U(\chi_M f), Ug) = \int_M fg^* \, d\mu$$

for every M in **S**. This means that $f(t)g^*(t) = 0$ for almost every t (with respect to the measure μ) and hence that there exists a set M in **S** such that $f(t) = 0$ for almost every t in M and $g(t) = 0$ for almost every t in $X - M$. For this set M we have

$$(E(M)x, x) = \| E(M)Uf \|^2 = \| U(\chi_M f) \|^2 = \int_M | f |^2 \, d\mu = 0$$

and similarly $(E(X - M)y, y) = \int_{X-M} | g |^2 \, d\mu = 0$, whence $\rho(x) \perp \rho(y)$ as asserted.

THEOREM 2. *If x and y are vectors, a necessary and sufficient condition that $\rho(x) \ll \rho(y)$ is that $C(x) \leqq C(y)$.*

Proof. Write $x = y_0 + z_0$ with y_0 in the range of $C(y)$ and z_0 orthogonal to the range of $C(y)$. Since $\rho(x) \equiv \rho(y_0) \vee \rho(z_0)$, it follows that if $\rho(x) \ll \rho(y)$, then $\rho(z_0) \ll \rho(y)$. Since, on the other hand, $C(y)\,C(z_0) = 0$, it follows from Theorem 1 that $\rho(z_0) \perp \rho(y)$. Consequently $\rho(z_0) = 0$, so that $z_0 = 0$, and therefore $x = y_0$. In other words x belongs to the range of $C(y)$, and therefore $C(x) \leqq C(y)$.

If, conversely, $C(x) \leqq C(y)$, we write $\mu = \rho(x)$ and $\nu = \rho(y)$. If $\nu(M) = 0$ for some M in $\mathbf{S}$, then $E(M)y = 0$ and consequently

$$E(M)C(y) = 0.$$

It follows that

$$E(M)C(x) = 0$$

and hence that

$$\mu(M) = (E(M)x, x) = \| E(M)x \|^2 = 0.$$

Our last result along these lines is of great technical significance; we call the reader's attention to the fact that, had we proved it in time, we could have used it to simplify slightly the proof of 63.1.

THEOREM 3. *If ν is a finite measure on $\mathbf{S}$ and if x is a vector in $\mathfrak{H}$ such that $\nu \ll \rho(x)$, then there exists a vector y in the range of $Z(x)$ such that $\nu = \rho(y)$; if $\nu \equiv \rho(x)$, then $Z(y) = Z(x)$.*

Proof. If $\mu = \rho(x)$, then, by the Radon-Nikodym theorem, there exists a non-negative function g in $\mathfrak{L}_1(\mu)$ such that $\nu(M) = \int_M g\, d\mu$ for every M in $\mathbf{S}$. If f is the non-negative square root of g, then $f \in \mathfrak{L}_2(\mu)$. If $y = Uf$, where U is the isomorphism described in 60.1, then

$$\nu(M) = \int_M |f|^2 d\mu = \| \chi_M f \|^2 = \| U(\chi_M f) \|^2$$
$$= \| E(M)Uf \|^2 = \| E(M)y \|^2.$$

If $\nu \equiv \rho(x)$, then, by Theorem 2, $C(x) = C(y)$. Since $Z(y) \leqq Z(x)$ and since $Z(x)$ is a row, it follows that $Z(y) = C(y)Z(x) = C(x)Z(x) = Z(x)$.

§66. Subspaces from Measures

THEOREM 1. *If μ is any finite measure on $\mathbf{S}$, then the set $\{x: \rho(x) \ll \mu\}$ is a subspace of $\mathfrak{H}$; if $C(\mu)$ is the projection on this subspace, then $C(\mu) \in \mathbf{F}$.*

Proof. If $\rho(x) \ll \mu$ and $\rho(y) \ll \mu$, then the relation

$$\| E(M)(\alpha x + \beta y) \| \leqq |\alpha| \cdot \| E(M)x \| + |\beta| \cdot \| E(M)y \|,$$

valid for all M in $\mathbf{S}$, shows that $\rho(\alpha x + \beta y)$ vanishes whenever both $\rho(x)$ and $\rho(y)$ vanish and hence whenever μ vanishes. If $\{x_n\}$ is a sequence of vectors such that $\rho(x_n) \ll \mu$ for all n and such that $x_n \to x$, then the relation $\| E(M)x_n \| \to \| E(M)x \|$ shows that $\rho(x)$ vanishes whenever all $\rho(x_n)$ vanish and hence whenever μ vanishes. It follows that

$$\{x: \rho(x) \ll \mu\}$$

is indeed a subspace and that, therefore, $C(\mu)$ may be defined. If $P \epsilon \mathbf{P}$ and if $\rho(x) \ll \mu$, then, since $P \leftrightarrow E$, it follows that

$$\| E(M)Px \| = \| PE(M)x \| \leqq \| E(M)x \|$$

for all M, and hence that $\rho(Px)$ vanishes whenever $\rho(x)$ vanishes. This implies that $\rho(Px) \ll \mu$ whenever $\rho(x) \ll \mu$, or, in other words, that P leaves invariant the range of $C(\mu)$. Consequently $P \leftrightarrow C(\mu)$ and therefore, since P is arbitrary, $C(\mu) \epsilon \mathbf{P}' = \mathbf{F}$.

THEOREM 2. *If μ is a finite measure on* S, *then $C(\mu)$ is separable; if $\mu = \rho(x)$, then $C(\mu) = C(x)$.*

Proof. Let R be a row such that $C(R) = C(\mu)$, and let $\{Z(x_j)\}$ be an orthogonal family of cycles such that $R = \vee_j Z(x_j)$. Since the fact that R is a row implies that $\{C(x_j)\}$ is an orthogonal family, and since $\rho(x_j) \ll \mu$ for all j, it follows from 65.1 that $x_j = 0$ except for countably many values of j. Since $C(\mu) = \vee_j C(x_j)$, it follows from 58.1 and 58.2 that $C(\mu)$ is separable. If $\mu = \rho(x)$ and if $\rho(y) \ll \mu$, then, by 65.2, $C(y) \leqq C(x)$, and consequently y belongs to the range of $C(x)$. In other words $C(\mu) \leqq C(x)$; the reverse inequality is obvious from the definition of $C(\mu)$.

THEOREM 3. *If μ and ν are finite measures on* S, *then*

$$C(\mu \wedge \nu) = C(\mu)C(\nu),$$

and therefore if $\nu \ll \mu$, then $C(\nu) \leqq C(\mu)$.

Proof. If $\rho(x) \ll \mu \wedge \nu$, then $\rho(x) \ll \mu$ and therefore x belongs to the range of $C(\mu)$. This implies that $C(\mu \wedge \nu) \leqq C(\mu)$. Since, similarly, $C(\mu \wedge \nu) \leqq C(\nu)$, it follows that $C(\mu \wedge \nu) \leqq C(\mu) C(\nu)$. If, on the other hand, x belongs to the range of $C(\mu) C(\nu)$, then $\rho(x) \ll \mu$ and $\rho(x) \ll \nu$, so that $\rho(x) \ll \mu \wedge \nu$. Since this means that x belongs to the range of $C(\mu \wedge \nu)$, it follows that $C(\mu) C(\nu) \leqq C(\mu \wedge \nu)$.

THEOREM 4. *If μ is a finite measure on X and x is a vector in $\mathfrak{H}$ such that $C(\mu)x = 0$, then $\mu \perp \rho(x)$.*

Proof. If $\rho(x) = \nu$, then, since $\nu \wedge \mu \ll \nu$, it follows from 65.3 that there exists a vector y in the range of $Z(x)$ such that $\rho(y) = \nu \wedge \mu$. Since $C(\mu)x = 0$, it follows that $C(\mu) Z(x) = 0$ and hence that $C(\mu)y = 0$. Since, however, $\rho(y) \ll \mu$, we know that y belongs to the range of $C(\mu)$. It follows that $y = 0$ and hence that $\mu \perp \nu$.

THEOREM 5. *If μ is a finite measure on* S *and if $\{\mu_j\}$ is a (necessarily countable) orthogonal family of finite measures on* S *such that $\vee_j \mu_j \equiv \mu$, then $C(\mu) = \vee_j C(\mu_j)$.*

Proof. Since Theorem 3 implies that $C(\mu) \geqq C(\mu_j)$ for all j, it is clear that $C(\mu) \geqq \bigvee_j C(\mu_j)$. Suppose, on the other hand, that x is any vector in the range of $C(\mu) - \bigvee_j C(\mu_j)$. Since x belongs to the range of $C(\mu)$, we have $\rho(x) \ll \mu$; since, at the same time, $C(\mu_j)x = 0$ for all j, it follows from Theorem 4 that $\rho(x) \perp \mu_j$ for all j and hence that $\rho(x) \perp \mu$. These two properties of $\rho(x)$ imply that $\rho(x) = 0$. We conclude that $x = 0$ and this completes the proof of the theorem.

§67. The Multiplicity Function of a Spectral Measure

If μ is a finite, non-zero measure on $\mathbf{S}$, the *multiplicity* of μ, in symbols $u(\mu)$, is defined to be the minimum value of the multiplicities $u(C(\nu_0))$ of the columns $C(\nu_0)$ determined by finite, non-zero measures ν_0 which are absolutely continuous with respect to μ; in other words

$$u(\mu) = \min \{u(C(\nu_0)) : 0 \neq \nu_0 \ll \mu\}.$$

If $\mu = 0$, we write $u(\mu) = 0$. We proceed quickly to show that the function u from measures to cardinal numbers is indeed a multiplicity function.

THEOREM 1. *If μ and ν are finite measures on* $\mathbf{S}$ *such that $0 \neq \nu \ll \mu$, then $u(\nu) \geqq u(\mu)$.*

Proof. If $0 \neq \nu_0 \ll \nu$, then $\nu_0 \ll \mu$ and therefore $u(\mu) \leqq u(C(\nu_0))$; since this inequality is valid for all admissible ν_0, it follows that

$$u(\mu) \leqq u(\nu).$$

THEOREM 2. *If $\{\mu_j\}$ is a countable orthogonal family of non-zero measures, and if $\mu \equiv \bigvee_j \mu_j$, then $u(\mu) = \min \{u(\mu_j)\}$.*

Proof. If $0 \neq \nu_0 \ll \mu$, and if $\nu_j \equiv \nu_0 \wedge \mu_j$ for each j, then $\nu_0 \equiv \bigvee_j \nu_j$. The last-written relation remains valid, of course, if the supremum is extended over only the set J of those values of j for which $\nu_j \neq 0$. It follows from 66.5 that $C(\nu_0) = \bigvee_{j \in J} C(\nu_j)$ and hence, from 64.2, that

$$u(C(\nu_0)) = \min_{j \in J} \{u(C(\nu_j))\} \geqq \min \{u(\mu_j)\}.$$

Since ν_0 is arbitrary, we see that $u(\mu) \geqq \min \{u(\mu_j)\}$. If, on the other hand, $0 \neq \nu_0 \ll \mu_j$ for some value of j, then $\nu_0 \ll \mu$ and therefore $u(C(\nu_0)) \geqq u(\mu)$, whence $u(\mu_j) \geqq u(\mu)$ for all j. This implies that

$$\min \{u(\mu_j)\} \geqq u(\mu).$$

The preceding two theorems tell us that the function u is a multiplicity function. We now have only one more technical detail to clear up before completing the theory, and that is the relation between the concepts of uniform multiplicity for measures and uniform multiplicity for projections.

THEOREM 3. *If x is a vector such that $C(x)$ has uniform multiplicity and if $\mu = \rho(x)$, then μ has uniform multiplicity. If, conversely, μ is a non-zero measure of uniform multiplicity and x_0 is a vector such that $C(\mu) = C(x_0)$ (cf. 66.2 and 58.3), then there exists a vector x such that (i) $\mu = \rho(x)$, (ii) $C(x)$ $(= C(\mu))$ has uniform multiplicity, and (iii) $Z(x) = Z(x_0)$.*

Proof. Suppose first that $\mu = \rho(x)$ and that $C(x)$ has uniform multiplicity. If $0 \neq \nu_0 \ll \mu$, then by 65.3 there exists a vector y_0 in the range of $Z(x)$ such that $\nu_0 = \rho(y_0)$. Since $y_0 \neq 0$, it follows that $C(\nu_0) \neq 0$. Since $C(\nu_0) \leqq C(\mu)$ by 66.3 and since $C(\mu) = C(x)$ by 66.2, it follows from the assumed uniformity that $u(C(\nu_0)) = u(C(\mu))$ (and that, therefore, $u(\mu) = u(C(\mu))$). If $0 \neq \nu_0 \ll \nu \ll \mu$, then, applying the result just proved, we obtain the relation $u(C(\nu_0)) = u(\mu)$, whence it follows that $u(\nu) = u(\mu)$. This proves the first assertion of the theorem.

To prove the second assertion, we suppose that μ has uniform multiplicity different from 0 and that x_0 is a vector such that $C(\mu) = C(x_0)$. It follows from this equation that $\rho(x_0) \ll \mu$; we propose to show that in fact $\rho(x_0) \equiv \mu$. For this purpose we let μ_0 be a measure (a relative complement of $\rho(x_0)$ in μ) such that $\mu_0 \perp \rho(x_0)$ and $\mu_0 \vee \rho(x_0) \equiv \mu$. If y_0 belongs to the range of $C(\mu_0)$, then $\rho(x_0) \perp \rho(y_0)$ and it follows from 65.1 that $C(x_0)C(y_0) = 0$, so that y_0 is orthogonal to the range of $C(x_0)$. The range of $C(x_0)$ is, however, the same as the range of $C(\mu)$, and, since $\mu_0 \ll \mu$, y_0 belongs to the range of $C(\mu)$. It follows that $y_0 = 0$, and therefore that $u(\mu_0) = 0$. Since, however, the assumed uniformity implies that if $\mu_0 \neq 0$, then $u(\mu_0) = u(\mu)$, it follows that $\mu_0 = 0$, and we do indeed have $\mu \equiv \rho(x_0)$. An application of 65.3 yields a vector x such that $Z(x) = Z(x_0)$ and $\rho(x) = \mu$ and hence, by 66.2, such that

$$C(x) = C(\mu) = C(x_0).$$

To prove that $C(x)$ has uniform multiplicity, suppose that F is a non-zero projection in **F** such that $F \leqq C(x)$. Since such an F is necessarily separable, there exists a vector y such that $F = C(y)$ and consequently, by 66.2, $F = C(\nu)$, where $\nu = \rho(y)$. The fact that $F \neq 0$ implies that $\nu \neq 0$. If $0 \neq \nu_0 \ll \nu$, then, by a repetition of a familiar argument, it follows that $C(\nu_0) \neq 0$ and therefore $u(C(\nu_0)) \geqq u(C(\nu))$, whence

$$u(\nu) \geqq u(C(\nu)).$$

Since, however, $u(\nu) = u(\mu)$ and $u(C(\nu)) \geqq u(\mu)$, it follows that

$$u(C(\nu)) = u(\mu).$$

Applying this result to μ in place of ν (i.e. letting $C(x)$ itself play the role of F), we see that $u(C(\nu)) = u(C(\mu))$; this completes the proof of the theorem.

§68. Conclusion

All the pieces are before us; all that remains is to put them together.

In the preceding section we have succeeded in associating a multiplicity function with every spectral measure. To the multiplicity function u we may apply 49.3 to obtain an orthogonal family $\{\mu_j\}$ of non-zero finite measures on **S** such that each μ_j has uniform multiplicity and such that $\mu \equiv \vee_j(\mu \wedge \mu_j)$ whenever μ is a finite measure on **S**. From 66.3 we see that $\{C(\mu_j)\}$ is an orthogonal family of projections in **F**; we assert that $\vee_j C(\mu_j) = 1$. If, indeed, x is an arbitrary vector in $\mathfrak{H}$ and if $\mu = \rho(x)$, then $\mu \equiv \vee_j(\mu \wedge \mu_j)$. It follows from 66.5 that

$$C(\mu) = \vee_j C(\mu \wedge \mu_j) \leqq \vee_j C(\mu_j),$$

and hence that the vector x belongs to the range of $\vee_j C(\mu_j)$. Since x is arbitrary, we may conclude that $\vee_j C(\mu_j) = 1$.

According to 67.3, for each fixed j, $C(\mu_j)$ has uniform multiplicity, and therefore, by 64.4, there exists an orthogonal family $\{R_{jk}\}$ of rows such that $C(\mu_j) = \vee_k R_{jk}$ and such that $C(R_{jk}) = C(\mu_j)$ for all k. The cardinal number of the index family $\{k\}$ is of course equal to $u(\mu_j)$. Since $C(\mu_j)$ is separable, it follows from 58.3 and 57.4 that each row R_{jk} is in fact a cycle. (The proof of this fact makes use of the elementary lemma which asserts that if x is a vector and R is a row such that

$$Z(x) \leqq R$$

and

$$C(x) = C(R),$$

then $Z(x) = R$.) Applying 67.3, we may find a family $\{x_{jk}\}$ of vectors (j is still fixed) such that $R_{jk} = Z(x_{jk})$ and such that $\rho(x_{jk}) = \mu_j$. By 60.1, the range of R_{jk} is isomorphic to $\mathfrak{L}_2(\mu_j)$ by an isomorphism which makes the given spectral measure E correspond to multiplications by the characteristic functions of measurable sets. Putting these isomorphisms together, first over all k, for fixed j, and then over all j, we obtain a representation of $\mathfrak{H}$ as a very large direct sum; each summand is the $\mathfrak{L}_2$ of a finite measure on **S**, and the representation makes correspond to the given spectral measure E the canonical spectral measure associated with $\{\mu_j\}$ and $\{u(\mu_j)\}$.

These considerations prove that every spectral measure is unitarily equivalent to a canonical one determined by its multiplicity function and hence that if two spectral measures have the same multiplicity function, they are unitarily equivalent. Suppose, conversely, that E and F are spectral measures with a common domain of definition and that U is a unitary operator such that $U^{-1}EU = F$. Write $\rho_E(x)$ for the measure μ defined by $\mu(M) = \|E(M)x\|^2$, and $\rho_F(x)$ for the measure μ defined by $\mu(M) = \|F(M)x\|^2$. If μ is any measure, if $\rho_F(x) \ll \mu$, and if $\mu(M) = 0$, then $\|E(M)Ux\| = \|U^{-1}E(M)Ux\| = \|F(M)x\| = 0$, whence $\rho_E(Ux) \ll \mu$. This means that if x belongs to the range of the projection which it is natural to denote by $C_F(\mu)$, then Ux belongs to the range of $C_E(\mu)$. Since, by symmetry, the converse is also true, we infer that $U^{-1}C_E(\mu)U = C_F(\mu)$. We may therefore conclude that the multiplicity associated with $C_E(\mu)$ via E is the same as the multiplicity associated with $C_F(\mu)$ via F, and hence that E and F have the same multiplicity function. This settles all our problems and fulfills all our promises.

REFERENCES

[The numbers in brackets refer to the bibliography that follows.]

For a discussion of vector spaces see [4: Chapter VII] or [6]. The point of view nearest to that of this book is to be found, of course, in [14].

The concept of a conjugate linear functional is closely related to the general algebraic concept of a semi-linear transformation; see, for instance, [19].

Occasionally one meets objects similar to inner product spaces but different in that the given bilinear functional is not required to be positive. For one such discussion, and for further references to the literature, see [43].

For an enlightening discussion of Schwarz's inequality, and many other inequalities useful in all parts of analysis, see [17].

An application of Schwarz's inequality for a positive (but not necessarily strictly positive) bilinear functional occurs in [29: p. 128].

Matrices such as $[\gamma_{jk}]$, mentioned at the end of §5, are called *Gramians*; for a further discussion see [7: Vol. I, p. 29].

The standard current sources of information about Hilbert space are [44] and [30].

Banach spaces are treated in [2] and [18]. Hille's book, incidentally, contains a discussion of operators on Banach spaces and of Banach algebras. Both subjects have many points of contact with this book.

The fact that Hilbert space is characterized among Banach spaces by its quadratic character is proved in [40]. Among the many other works dealing with various characterizations of Hilbert space, I explicitly mention four; the interested reader will find further references in their bibliographies. The four, in chronological order, are [20], [26], [28], and [8].

Nagy [30: p. 4] gives a very brief sketch of the reasons for the completability of an inner product space. A detailed discussion appears in [36: Vol. II, p. 13, or second printing, Vol. II, p. 17].

For a discussion of not necessarily countable sums of complex numbers see [37].

The fact that the Fourier coefficients of a vector with respect to an orthonormal sequence approach zero is an analog of a somewhat deeper analytic fact known as the Riemann-Lebesgue lemma; cf. [5: p. 3].

A detailed proof that some of the examples of Hilbert spaces described in §9 do indeed satisfy the axioms is given by Stone, [44: pp. 14–15 and pp. 23–32]. The proof of the completeness of the relevant function spaces appears also in [15: pp. 107 and 177].

The lattice-theoretic concepts relevant to the study of subspaces of a Hilbert space, as well as all others used in the sequel, are exhaustively treated in [3].

For a discussion of several analytic applications of the Gram-Schmidt orthogonalization process, and, in particular, of the polynomials mentioned in §14, see [47].

The completeness of the exponential functions in $\mathfrak{L}_2$ over a finite interval is a classical fact; see, for instance, [52: Vol. I, p. 11].

The example of a non-closed vector sum described in §15 is a slight paraphrase of an example given by Stone, [44: pp. 21–22]. A necessary and sufficient condition that the vector sum of two subspaces (of any Banach space) be closed is given in [23]. A simpler example of a non-closed vector sum which, because of its dependence on the concept of an operator, could not be given in §15, can be obtained as follows. Let A be an operator on a Hilbert space $\mathfrak{H}$, such that the range $\mathfrak{R}$ of A is not closed. In the direct sum of $\mathfrak{H}$ with itself, let $\mathfrak{M}$ be the set of all vectors of the form $\{x, 0\}$ and let $\mathfrak{N}$ be the set of all vectors of the form $\{x, Ax\}$ with $x \in \mathfrak{H}$. It is easy to verify that both $\mathfrak{M}$ and $\mathfrak{N}$ are subspaces of the direct sum and that $\{x, y\} \in \mathfrak{M} + \mathfrak{N}$ if and only if $y \in \mathfrak{R}$; since $\mathfrak{R}$ is not closed, neither is $\mathfrak{M} + \mathfrak{N}$.

The elegant proof (in §16) of the uniqueness of dimension in the infinite case is due to Löwig, [27]. The corresponding theorem for finite-dimensional spaces is settled, for instance, in [4: p. 179].

J. v. Neumann, [35], discusses the structure of integral operators such as the ones mentioned in §20(i). The matrix point of view is treated in [51].

The discussion of an operator which is not reduced by any non-trivial subspace could easily have been much more complicated. The elegant trick in §23 was shown to me by W. A. Howard.

The definition of summability for a family of operators (§28) is as close as this book ever gets to mentioning the various well-known topologies for operators. Their importance was first emphasized by von Neumann, [34]; more recently the subject has been discussed by Dixmier, [9].

The range of the infimum of two not necessarily commutative projections is determined in [36: Vol. II, p. 38, or second printing, Vol. II, p. 55].

The proof of the fact that the norm of a Hermitian operator A can be calculated from its spectrum is usually based on the relation

$$\| A \| = \sup \{ | (Ax, x) | : \| x \| = 1 \}.$$

The more elementary proof in §34 was called to my attention by W. A. Howard. If a similar proof could be constructed of the corresponding theorem for normal operators (or of the characterization of $\| A \|$ by the values of (Ax, x) on the unit sphere), the proof of the spectral theorem for normal operators could be immensely simplified.

The fact that every operator has a non-empty spectrum is proved in [44: p. 149]. For a related, abstract discussion see [12].

The heuristic discussion in §35 is strongly influenced by the existence of general theories which include, as special cases, the pertinent part of the theory of real functions as well as of spectral theory; see, for example, [45], [31], [42], and [21].

Measurable spaces are defined and discussed in [15: p. 73, et seq.].

The juggling with differentials which occurs in §37 and several other places is justified in [15: pp. 132–134].

The concept of regularity for spectral measures imitates a well-known numerical concept; cf., for instance, [15: Chapter X]. It should be remarked that the class of all Borel sets in a locally compact Hausdorff space (as defined in [15]) is not necessarily a σ-algebra (but merely a σ-ring). The phrase "Borel set" is used in this book in its classical meaning as an element of the σ-algebra generated by the class of all open sets. An important example of regular spectral measures in locally compact topological groups (to which the theory in Chapter III of this book is applicable) occurs in [1].

The fact that every finite measure on the class of all Borel subsets of the complex plane is regular follows from [15: Theorem E, p. 218 and Theorem G, p. 228].

The proof of 40.1 makes use of the uniqueness assertion of the representation theorem stated in §0. Though this constitutes only the first and weakest use of that theorem, it provides a good excuse for mentioning the relevant literature. It is customary to break up the theorem into two parts, one of which (the Weierstrass approximation theorem) asserts that continuous functions can be approximated by polynomials, and the other of which (the Riesz representation theorem) asserts the existence of a suitable complex measure associated with a bounded linear functional of continuous functions. The one-dimensional Weierstrass theorem is treated in many standard texts; cf., for instance, [50: p. 152]. An amusing interpretation of the proof in the language of probability theory is given in [48: p. 116]. The Weierstrass theorem for any finite number of dimensions is proved in [13: p. 123]. A general and modern discussion of the circle of ideas centering around the Weierstrass theorem appears in [46]. A proof of the Riesz representation theorem for the general case can be dug out of [15: Theorem D, p. 247 and Exercise 5, p. 249]. Kakutani, [22: p. 1012], gives a statement which is more readily applicable to present needs, but for real-valued linear functionals only; the complex case is, however, a trivial corollary which follows simply from the consideration of the real and imaginary parts of the given linear functional.

The problem discussed in §§41 and 42 may (in virtue of 40.1) be stated as follows: If E is a compact, complex spectral measure, if $A = \int \lambda dE(\lambda)$, and if B is an operator such that $A \leftrightarrow B$, then does it or does it not follow that $A^* \leftrightarrow B$? Since A is normal, the question may also be viewed as a special case of the problem of transitivity of commutativity: if $A^* \leftrightarrow A$ and $A \leftrightarrow B$, can one infer that $A^* \leftrightarrow B$? The problem was explicitly raised more than ten years ago by von Neumann; it appeared in print in [39]. The first solution is due to Fuglede, [11]; the solution presented in §§41 and 42 appears in [16].

The neat and powerful characterization of spectral subspaces (41.1) was proved for Hermitian operators in [24].

The neat arrangement of the ideas in the proof of the spectral theorem for Hermitian operators, as given in §43, is due to Eberlein, [10].

The crucial measure-theoretic extension theorem needed for the proof of the spectral theorem for normal operators in §44 may be found in [36: Vol. I, p. 146, or second printing, Vol. I, p. 167].

The Radon-Nikodym theorem is standard measure-theoretic equipment; cf., for instance, [15: p. 128]. A very neat proof based almost exclusively on geometric facts about Hilbert space occurs in [38: p. 127].

The concept of a multiplicity function appears explicitly in [41]. The first successful attempt to construct a theory of multiplicities for non-separable Hilbert spaces was made by Wecken, [49]. The theory for separable Hilbert spaces is presented by Stone, [44: Chapter VII], who also gives references to the classical literature and, in particular, to Hellinger's original solution of the problem of unitary equivalence.

The "prime" operation described in §53 is inspired by [34: pp. 388–389].

The representation theorem for Boolean algebras which is mentioned in §54 can be found in [25]. It is worth noting that the conditions that the relevant representation theorem requires of the Boolean algebra **F** are much weaker than the ones that come free with the **F** in the text; all that is necessary is that **F** be a σ-algebra. Another proof of the representation theorem, closer in spirit to Stone's topological approach, is outlined in [15: Exercise 15c, p. 171].

The term "separable" as used in §58 is due to Nakano, [33]. The work of Nakano, as represented by this paper and an earlier one, [32], is one of the main sources on which the exposition in Chapter III is based.

BIBLIOGRAPHY

[1] AMBROSE, W., *Spectral Resolution of Groups of Unitary Operators*. Duke Math. J. 11 (1944), 589–595.

[2] BANACH, S., *Théorie des Opérations Linéaires*. Warsaw, 1932. [*reprint*, New York, 1943].

[3] BIRKHOFF, G., *Lattice Theory*. New York, 1948.

[4] BIRKHOFF, G., and S. MACLANE, *A Survey of Modern Algebra*. New York, 1948.

[5] BOCHNER, S., and K. CHANDRASEKHARAN, *Fourier Transforms*. Princeton, 1949.

[6] BOURBAKI, N., *Éléments de Mathématique, Livre II: Algèbre, Chapitre II: Algèbre Linéaire*. Paris, 1947.

[7] Courant, R., and D. Hilbert, *Methoden der Mathematischen Physik.* Berlin, 1931. 2 Vols. [Vol. I, trans., New York, 1953; Vol. II, *reprint,* New York, 1943].

[8] Day, M. M., *Some Characterizations of Inner-product Spaces.* Trans. A. M. S. 62 (1947), 320–337.

[9] Dixmier, J., *Les Fonctionnelles Linéaires sur l'Ensemble des Opérateurs Bornés d'un Espace de Hilbert,* Ann. Math. 51 (1950), 387–408.

[10] Eberlein, W. F., *A Note on the Spectral Theorem,* Bull. A. M. S. 52 (1946), 328–331.

[11] Fuglede, B., *A Commutativity Theorem for Normal Operators,* Proc. N. A. S. 36 (1950), 35–40.

[12] Gelfand, I., *Normierte Ringe,* Mat. Sbornik 9 (1941), 3–24.

[13] Graves, L. M., *The Theory of Functions of Real Variables.* New York, 1946.

[14] Halmos, P. R., *Finite Dimensional Vector Spaces.* Princeton, 1958.

[15] Halmos, P. R., *Measure Theory.* New York, 1950.

[16] Halmos, P. R., *Commutativity and Spectral Properties of Normal Operators,* Acta Szeged 12 (1950), 153–156.

[17] Hardy, G. H., J. E. Littlewood, and G. Pólya, *Inequalities.* Cambridge, 1952.

[18] Hille, E., and R.S. Phillips, *Functional Analysis and Semi-groups.* 1957.

[19] Jacobson, N., *The Theory of Rings.* New York, 1943.

[20] Kakutani, S., *Some Characterizations of Euclidean Space,* Jap. J. Math. 16 (1939), 93–97.

[21] Kakutani, S., *Concrete Representation of Abstract (L)-spaces and the Mean Ergodic Theorem,* Ann. Math. 42 (1941), 523–537.

[22] Kakutani, S., *Concrete Representation of Abstract (M)-spaces. (A characterization of the Space of Continuous Functions.)* Ann. Math. 42 (1941), 994–1024.

[23] Kober, H., *A Theorem on Banach Spaces,* Comp. Math. 7 (1939), 135–140.

[24] Lengyel, B. A., and M. H. Stone, *Elementary Proof of the Spectral Theorem,* Ann. Math. 37 (1936), 853–864.

[25] Loomis, L. H., *On the Representation of σ-complete Boolean Algebras,* Bull. A. M. S. 53 (1947), 757–760.

[26] Lorch, E. R., *The Cauchy-Schwarz Inequality and Self-adjoint Spaces,* Ann. Math. 46 (1945), 468–473.

[27] Löwig, H., *Über die Dimension Linearer Räume,* Studia Math. 5 (1934), 18–23.

[28] Mackey, G. W., and S. Kakutani, *Ring and Lattice Characterizations of Complex Hilbert Space,* Bull. A. M. S. 52 (1946), 727–733.

[29] Murray, F. J., and J. v. Neumann, *On Rings of Operators,* Ann. Math. 37 (1936), 116–229.

[30] Nagy, B. v. Sz., *Spektraldarstellung Linearer Transformationen des Hilbertschen Raumes.* Berlin, 1942.

[31] Nakano, H., *Teilweise Geordnete Algebra,* Jap. J. Math. 17 (1940), 425–511.

[32] Nakano, H., *Unitärinvariante Hypermaximale Normale Operatoren,* Ann. Math. 42 (1941), 657–664.

[33] Nakano, H., *Unitärinvarianten im Allgemeinen Euklidischen Raum*, Math. Ann. 118 (1941), 112–133.

[34] Neumann, J. v., *Zur Algebra der Funktionaloperationen und Theorie der Normalen Operatoren*, Math. Ann. 102 (1930), 370–427.

[35] Neumann, J. v., *Charakterisierung des Spektrums Eines Integraloperators*. Paris, 1935.

[36] Neumann, J. v., *Functional Operators*. Princeton, 1935; second printing, Princeton, 1950.

[37] Neumann, J. v., *On Infinite Direct Products*, Comp. Math. 6 (1938), 1–77.

[38] Neumann, J. v., *On Rings of Operators*, III, Ann. Math. 41 (1940), 94–161.

[39] Neumann, J. v., *Approximative Properties of Matrices of High Finite Order*, Port. Math. 3 (1942), 1–62.

[40] Neumann, J. v., and P. Jordan, *On Inner Products in Linear, Metric Spaces*, Ann. Math. 36 (1935), 719–723.

[41] Plessner, A. I., and V. A. Rohlin, *Spectral Theory of Linear Operators*, II, Uspehi Mat. Nauk 1 (1946), No. 1, 71–191.

[42] Riesz, F., *Sur Quelques Notions Fondamentales dans la Théorie Générale des Opérations Linéaires*, Ann. Math. 41 (1940), 174–206.

[43] Savage, L. J., *The Application of Vectorial Methods to Metric Geometry*, Duke Math. J. 13 (1946), 521–528.

[44] Stone, M. H., *Linear Transformations in Hilbert Space and their Applications to Analysis*. New York, 1932.

[45] Stone, M. H., *A General Theory of Spectra*, I, Proc. N. A. S. 26 (1940), 280–283; II, Proc. N. A. S. 27 (1941), 83–87.

[46] Stone, M. H., *The Generalized Weierstrass Approximation Theorem*, Math. Mag. 21 (1948), 167–184, 237–254.

[47] Szegö, G., *Orthogonal Polynomials*. New York, 1939.

[48] Uspensky, J. V., *Introduction to Mathematical Probability*. New York, 1937.

[49] Wecken, F., *Unitärinvarianten Selbstadjungierter Operatoren*, Math. Ann. 116 (1939), 422–455.

[50] Widder, D. V., *The Laplace Transform*. Princeton, 1941.

[51] Wintner, A., *Spektraltheorie der Unendlichen Matrizen*. Leipzig, 1929.

[52] Zygmund, A., *Trigonometrical Series*. 2 Vols. Cambridge, 1960.